A Strategic Nature

A Strategic Nature

*Public Relations and the Politics
of American Environmentalism*

MELISSA ARONCZYK AND MARIA I. ESPINOZA

OXFORD
UNIVERSITY PRESS

OXFORD
UNIVERSITY PRESS

Oxford University Press is a department of the University of Oxford. It furthers
the University's objective of excellence in research, scholarship, and education
by publishing worldwide. Oxford is a registered trade mark of Oxford University
Press in the UK and certain other countries.

Published in the United States of America by Oxford University Press
198 Madison Avenue, New York, NY 10016, United States of America.

CIP data is on file at the Library of Congress

ISBN 978-0-19-005535-6 (pbk.)
ISBN 978-0-19-005534-9 (hbk.)

DOI: 10.1093/oso/9780190055349.001.0001

Contents

Figures and Tables

Figures

Tables

Abbreviations

ADR	Alternative Dispute Resolution
AEF	Andersson Elffers Felix Public Affairs
AFL	American Federation of Labor
AISI	American Iron and Steel Institute
CAA	Clean Air Act (U.S.)
CF&I	Colorado Fuel & Iron Company
CIIT	Chemical Industry Institute of Toxicology
CIO	Congress of Industrial Organizations (U.S.)
CMA	Chemical Manufacturers Association (see MCA)
CSI	Clean Sites, Inc.
D4CA	Data for Climate Action
EDF	Environmental Defense Fund
EIS	Environmental Information Systems
EPA	Environmental Protection Agency (U.S.)
GEMI	Global Environmental Management Initiative
ICC	International Chamber of Commerce
IEB	International Environmental Bureau
IPRA	International Public Relations Association
MCA	Manufacturing Chemists' Association (founded 1872; renamed Chemical Manufacturers Association in 1978; renamed American Chemistry Council in 2000)
NAFTA	North American Free Trade Agreement
NAM	National Association of Manufacturers (U.S.)
NASA	National Aeronautics and Space Administration (U.S.)
NCPP	National Coal Policy Project (U.S.)
NEDA	National Environmental Development Association (U.S.)
NEDA/CAAP	National Environmental Development Association—Clean Air Act Project
NEDA/CWP	National Environmental Development Association—Clean Water Project
NIRA	National Industrial Recovery Act (U.S.)
NRDC	Natural Resources Defense Council
PBS	Public Broadcasting Service (U.S.)
PCEQ	President's Commission on Environmental Quality (U.S.)
PRSA	Public Relations Society of America
SDG	Sustainable Development Goals (United Nations)

SIPI   Scientists' Institute for Public Information
SWOC   Steel Workers Organizing Committee
UNCED   United Nations Conference on Environment and Development
UNCHE   United Nations Conference on the Human Environment
UNEP   United Nations Environment Programme
USCIB   United States Council for International Business
US-ICC   U.S. Interstate Commerce Commission
WEF   World Economic Forum
WICEM   World Industry Conference on Environmental Management

Acknowledgments

This is a book about the conditions of knowledge in the making of the natural environment as a public problem. To ask what we know and how we know it is an inherently social project. It relies in every instance on the willingness of others to share their own knowledge and understanding. It is in this spirit of inquiry that we acknowledge with deep gratitude the "public" that came into existence around this book.

To Lee Edwards, Jeff Pooley, Devon Powers, Thomas Rudel, Chris Russill, and Tim Wood, who read the entire manuscript and provided such thoughtful and generous comments. To Monika Krause, for support and perspective. To Bob Brulle, for his mentorship and guidance along the twists and turns of the research path. To Jeff Alexander, Philip Smith, Fred Wherry, and the members of the Center for Cultural Sociology at Yale University, who listened to and helped shape the earliest stirrings of this project.

To the librarians and archivists at the Rutgers and Yale libraries, the Wisconsin Historical Society, and the Smithsonian Institution Archives for their expertise; to Jamie Corey, Mahogany Lore, and Jayde Valosin for dedicated research assistance; to James Cook, Emily Mackenzie, Jeremy Toynbee, and Patterson Lamb for their exceptional editorial guidance; and to Kimberly Glyder for her stunning cover design.

To the many PR people, lobbyists, consultants, administrative and political officials, communications managers, industry representatives, nonprofit leaders, environmental advocacy groups, government agencies, activists, organizers, and media companies who allowed us to speak with and learn from them. To our families, who supported this project in more ways than they know.

To the National Science Foundation, whose funding made this project possible; and the School of Communication & Information and the Department of Sociology at Rutgers University for giving us additional resources, including the gift of time, to support this work.

Perhaps our greatest debt is owed to the public relations counselor E. Bruce Harrison (1932–2021). It is a complex debt, and not one that can be easily untangled. Harrison was a brilliant strategist who worked with hundreds of industrial and political leaders as well as journalists and media organizations from the 1960s through the early 2000s. Starting in February, 2017, after his immediate response to a tentative email request, we maintained a semi-regular correspondence until ill health befell him in 2020. In our visits together, our lunches, his introductions to old friends and allies, his emails and letters, he was unfailingly gracious and charming. His decision to share his company library with us—a total of over 850 documents spanning his fifty-year career—was at once an act of tremendous generosity and a personal desire to be remembered for his role in what he called "sustainable communication"—the "greening" of industry through expert PR. Without him, this book could not have been written.

But without him, we also wonder whether the contemporary crisis of global warming would have become so dire. Advancing communication, rather than environmental action, as the locus of sustainability, Harrison showed relentless determination to protect and promote his clients—major polluters and environmental rule-breakers in the most contentious industries in the world, from oil and coal to tobacco and pesticides. His work has contributed in no small way to the misplaced belief in private sector strategies to "solve" the climate crisis despite daily indications of the failure of these strategies to do much besides promote themselves.

Of course, Harrison could not act alone. As we hope this book will show, any reckoning with environmentalism in America must account for the long-term political, economic, and information conditions in which public relations was able to become such a powerful force. This book is dedicated to all those working to turn the tides against the easy acceptance of clearly unacceptable practices, so that we may imagine more considered and healthy futures.

Introduction
Public Relations and Its Problems

On a wet spring day in 2019, dogwood and magnolia trees in brilliant bloom, a group of media makers, environmental activists and communications professionals gathered at the Columbia School of Journalism in New York City for the launch of Covering Climate Change: A New Playbook for a 1.5-degree World. Hosted by the progressive magazines *The Nation* and *The Columbia Journalism Review,* with media sponsors *The Guardian* and New York City public radio station WNYC, the goal of the event was to "begin a conversation that America's journalists and news organizations must have with one another, as well as with the public we are supposed to be serving, about how to cover this rapidly uncoiling emergency."[1]

The first speaker, Bill McKibben, was well known to the attendees. Former staff writer at the *New Yorker*, founder of the grassroots environmental activist organization 350.org, his 1989 book, *The End of Nature*, is broadly credited with turning global warming into a public problem. Thirty years and multiple books and honors later, McKibben is heralded as an influential organizer dedicated to provoking public action against polluting industries and their political and economic support systems.

Speaking by Skype ("we're learning to use these low carbon technologies"), McKibben began by putting his finger squarely on the pulse of the thirty-year problem preventing the public from knowing the truth about climate change:

> We know now much more of the behind-the-scenes story than we did even a few years ago. . . . Beginning right after [NASA scientist] Jim Hansen testified to Congress [in 1988], the oil industry began the project—with the utility industry and the coal industry—of setting up a kind of architecture of denial and misinformation. And the strategy they hit on was the same strategy that the tobacco industry had hit on—indeed they hired many of the veterans of that industry—and that strategy was to try and pretend that there was doubt about the situation.
>
> And climate change was new enough as a topic at first that it was a fairly plausible strategy. For a few years as scientists were kind of getting their

A Strategic Nature. Melissa Aronczyk and Maria I. Espinoza, Oxford University Press. © Oxford University Press 2022.
DOI: 10.1093/oso/9780190055349.003.0001

ducks in a row, it's understandable that journalism fell for the creation of what was in essence a phony debate. The strategy of the industry and its PR teams was to insist that we didn't know if global warming was real. . . . And the phoniness of this debate is that both sides knew the answer to that question right at the beginning. It's just that one of them was willing to mount a PR offensive in the opposite direction of the truth. And that PR offensive was obviously extremely successful. . . .

In other words, this was one of the cases where the PR guys . . . got the better of us for a very long time. And that was tragic, because the three decades essentially that we wasted in this phony debate were the three decades that we most needed in order to come to terms with climate change.[2]

The story of "PR guys" winning the war of information around the causes of global warming is well supported by scholarly research and investigative reporting.[3] These accounts have brought to light many of the devious information strategies by which fossil fuel industries cast doubt on scientific knowledge.

But McKibben's story serves another function, which gets to the heart of what this book is about. It presents the lack of media coverage of climate change as a problem of bad information. It sees the failure in the media's poor publicity, giving rise to "a calamitous public ignorance."[4] And it sees the solution at least partly in the media's responsibility to overcome the distortions of false and self-interested information to provide the public with the truth of scientific facts.

The problem with this critique is that it both overplays and underplays the explanatory power of public relations as a system of influence in our information-mediated lives. It overplays the role of PR in the lack of public responses to climate change while underplaying its role in damaging the vital relationship between people and the natural environment. It overplays the opposition between journalism and PR, presenting these professions as harbingers of truth and lies, respectively, while limiting PR to a system of messaging and framing. It overplays the authority and responsibility of journalists themselves in enacting behavioral change among its publics while downplaying the authority and responsibility of PR agents, characterizing them as mere mouthpieces for their powerful clients.

Those who embody this critique know that the story is more complex than it appears. But even when we add in some of the deeply embedded structural and political factors that have contributed to the lack of action around climate change in the United States, these basic premises about PR hold fast. In the making of public knowledge about climate change, PR is understood as having three core characteristics. It is defined as spin, that is, information that obstructs, manufactures, or manipulates facts about environmental problems; it is seen as a handmaiden to industrial power, amplifying anti-environmental strategies designed in corporate boardrooms; and it is perceived to be a source of public cynicism and disaffection, the "bad other" to journalism's moral rectitude.

This book offers another way to think about public relations. It examines the roots of these perceptions in order to present a more robust account of the ways that strategic communication—and its communicators—have wielded influence over the relationship among information, the environment, and its publics in a modern democracy. It demonstrates how public relations specialists actively construct and manage public understandings of the environment. It shows the mechanics of environmental publicity, bringing front and center what is so often characterized as the behind-the-scenes work of PR. To do this, we rely on both macro- and micro-level investigations, combining insights from national patterns and individual motivations to develop a conceptual framework for understanding the promotional culture around the communication of the environment.

To draw the big picture of the impact of promotional culture on environmental thinking, we adopt a historical perspective. The relationship of information, environment, and publicity is a long-standing one. Throughout the twentieth century, making the natural environment into a matter of public concern required a series of techniques of mediation. Strategic communicators from as early as the Progressive Era made use of publicity techniques to shape both environmental awareness and the political landscape on which this awareness could take root. These mediations shaped how environmental problems were thought about, given credence, or dismissed. The central argument in this book is that American environmentalism emerged alongside the tools, techniques, and expertise of American public relations, and that neither environmentalism nor PR would look the way it does today without the other. Of course, these concepts did not evolve in a vacuum. Understanding the relationship of public relations to environmentalism requires a focus on the simultaneously evolving systems of mass media and public opinion, environmental regulation and legislation, and the creation and circulation of information about environmental issues.

The basic premise of this book is that it is not possible to understand the role of the environment in our everyday lives without understanding how something called "the environment" has been invented and communicated to us throughout our lives. To tell this story properly requires a careful account of the evolution of the institutions, norms, and movements that have pushed environmental concerns to the fore of public opinion and political action. But it also demands an examination of the simultaneous evolution of professional communicators and the formation of *their* institutions, norms, and movements. Without this piece of the puzzle, we miss crucial ways that struggles are won, resources allocated, and beliefs fostered about environmental problems.

The Historical Roots of Publicity

McKibben's story reproduces an enduring American anxiety over the role of publicity in the making of informed publics in a democracy. The history of the concept of publicity reveals some of this ambivalence. In 1926, the philosopher John Dewey advanced an idea that would become paramount to American democratic thought: that "there can be no public without full publicity in respect to all consequences which concern it." For people to recognize themselves as members of a public, with the power to pronounce on matters of social importance, these matters must be "observed, reported and organized" through "free and systematic communication." One of the problems confronting the democratic organization of publics, in Dewey's eyes, lay in the "physical agencies of publicity," from advertising and propaganda firms to sensationalist news outfits. In the newly industrialized and technological post–World War I age, these "exploiters of sentiment and opinion" threatened to eclipse the possibility of public congress. If societies "demand communication as a prerequisite" for participation in shared interests and institutions, the use of communication to manipulate public feeling or provoke cheap responses to unworthy issues precluded the possibility of forming this shared outlook.[5]

Dewey's contemporary, the political theorist Walter Lippmann, was equally concerned about the role of publicity in the making of democratic publics. His solutions to the crisis of public discourse were rooted in the discipline of scientific reason. In *Public Opinion* (1922), Lippmann laments the sheer complexity of political and social affairs, the failings of modern communications media, and the fragmentation of attention, all of which limited citizens' ability to know what they needed to know to make good decisions about how society should operate. Living in a "world beyond our reach," Americans are subjected to the "manufacture of consent"—the manipulation of public sentiment through the "self-conscious art" of professional persuasion, and its troubling legacy is that "the knowledge of how to create consent will alter every political calculation and modify every political premise." The ideal of the "omnicompetent" citizen who could be informed on all issues of public importance was not possible under these conditions, where "the practice of appealing to the public on all sorts of intricate matters means almost always a desire to escape criticism from those who know by enlisting a large majority which has had no chance to know."[6]

It is hard to write a book about democratic communication in the American public sphere without drawing on the ideas of Dewey and Lippmann. The ideals they advanced continue to test our evolving values and beliefs about the role of publicity in fostering the unity of public purpose; and the concerns they raised about the fetters on this purpose continue to challenge us in our assessment of our social and political institutions.[7]

There is a third figure from that era, less often cited in this context, but one whose ideas have arguably become just as central to our understanding of the role of publicity in the making of informed publics: the public relations counselor Edward Bernays.

Bernays is generally known as the "father" of public relations, a title that he may well have invented for himself. For the better part of the twentieth century, Bernays was devoted to inventing, legitimating, and advancing the profession of PR. He taught its first academic course; published dozens of treatises; and developed hundreds of promotional campaigns for clients of all stripes, from the American Tobacco Company to civic organizations.

But the most important contribution Bernays made to the concept of the public in the burgeoning democratic life of the early twentieth century is also its most contested: the transformation of the concept into a strategic resource.

In 1923, a year after Lippmann's *Public Opinion* was published, Bernays hastily put out his own missive, *Crystallizing Public Opinion*. Bernays wanted to transpose Lippmann's concerns about the machinery of publicity and "the manufacture of consent" into his terms, advancing PR as a necessary feature of democracy rather than a fetter on it. Indeed, the primary objective of

Crystallizing Public Opinion was to promote public relations as an invaluable profession for the exercise of democracy in the modern era.

To turn his business into a credible input to democracy, Bernays sought to move public opinion from the realm of normative democratic theory to the practice of expert technical management. Bernays celebrated what Lippmann (and later, political theorists Habermas and Bourdieu) decried: the use of media, polling, surveys, and other techniques to make and manage publics.[8] He drew liberally and selectively on expert sources from across the social sciences to lend his ideas an air of respectability.[9] He often invoked his family connection to the psychoanalyst Sigmund Freud (he was Freud's nephew). At New York University he taught the first course on public relations ever offered, using his entrée into the academic community to codify his ideas. He advocated the use of legal metaphors to describe PR practice, referring to himself as a PR counselor, whose business was conducted in the court of public opinion.

Writing in the *American Journal of Sociology* in 1928, Bernays advanced the idea of the public opinion specialist, demonstrating how sociological theories and methods could be made useful to the technical process of public manipulation. Armed with an understanding of group dynamics, statistics, and the impact of affect on behavior, the analyst

> has methods adapted to educating the public to new ideas, to articulating minority ideas and strengthening them, to making latent majority ideas active, to making an old principle apply to a new idea, to substituting ideas by changing clichés, to overcoming prejudices, to making a part stand for the whole, and to creating events and circumstances that stand for his ideas. He must know the physical organs of approach to his public: the radio, the lecture platform, the motion picture, the letter, the advertisement, the pamphlet, the newspaper. He must know how an idea can be translated into terms that fit any given form of communication, and that his public can understand.[10]

What Bernays understood deeply was the power of "the public" as a cultural form. In his eyes, the public could be invoked as a strategic resource—not as an end in itself, but as a means for other ends. If you retain the ideals of publicity as a principle of democracy, consensus as the desired outcome of reasonable (in Habermas's terms, "rational-critical") debate, and communication as the feeder for social integration, you can conduct your affairs in the name of the public good. Whether advancing the public good was your actual motivation became less important than maintaining the ideals that surrounded it.

The scholar Sue Curry Jansen has shown how Bernays consistently applied what he favorably called "semantic tyranny" to existing phrases and concepts, retooling them into ideas he could use for his emerging PR practice. The phrase, "manufacturing consent," for instance, initially uttered by Walter Lippmann as a damning critique of propaganda, became in Bernays's hands a desirable objective. More damaging still, Jansen argues, was the way that Bernays made Lippmann himself into an apologist for elite expertise as the source of effective government, a misinterpretation that persists in various forms today.[11]

Bernays's signal accomplishment was to advance and institutionalize the notion that ideas and information need to be shaped, framed, and labeled in order to appear acceptable to people in political and cultural contexts. For Bernays, PR was less about communication than it was about creating contexts for communication: contexts where certain ideas and information could be made to seem relevant and legitimate while others receded or became marginal. By seating ideas and information within democratic structures of participation, communication, and social importance, publics could be formed and opinions garnered around the issues of the day. This is the foundational definition of public relations—creating relational meanings to structure groups of people who are enjoined to think of themselves as legitimate publics in a democracy.

This capsule history of Lippmann, Dewey, and Bernays is meant to serve a double function. In describing how the cultural form of the public can be adapted to serve strategic and self-interested ends, the story also shows how social and political thought and action can themselves adapt to these transformed ends. Whether we accept or lament them, the industries of public relations and public opinion are part of our modern democratic system today. These industries belong at the forefront of our thinking about democratic publics—not to celebrate them, but to recognize how central their work has been to modern understanding about public communication. This recognition paves the way for social and political researchers to develop sharper tools of critique.

Most critics dismiss PR as manipulative distortions of reality. In the face of mounting evidence of organized misinformation campaigns financed by fossil fuel industries to deny and obfuscate global warming, this is unquestionably the case. But the question at the core of this book is how such manipulations have been so devastatingly effective. It is hard to ignore the will to mislead, especially when those doing the misleading are armed with considerable authority or resources. There are sometimes reasons beyond resource differentials, however, that these misinterpretations are deemed acceptable as declarative statements of how things are.

Arguments about manipulated publicity are not as helpful as they need to be to understand how PR operates. When PR is taken as manipulation, this activates the premise that we render the world transparent by bringing this manipulation to light, revealing the "truth" underneath.[12] Yet as recent political events have clearly demonstrated, the opposition of manipulation (fakery, distortion, lies) to truth (honesty, transparency, facts) reproduces an unreflexive antinomy that is neither analytically nor rhetorically sustainable.

Hannah Arendt argued that the problem of politics and truth was the central dilemma of the twentieth century. "Seen from the viewpoint of politics," she wrote, "truth has a despotic character":

> Facts are beyond agreement and consent, and all talk about them—all exchanges of opinion based on correct information—will contribute nothing to their establishment. Unwelcome opinion can be argued with, rejected, or compromised upon, but unwelcome facts possess an infuriating stubbornness that nothing can move except plain lies. The trouble is that factual truth, like all other truth, peremptorily claims to be acknowledged and precludes debate, and debate constitutes the very essence of political life.[13]

Let us return to McKibben to bear this out. McKibben argues that the appearance of debate in our media is what has prevented the public from coming to terms with climate change. But the appearance of debate is not the problem. The problem is that we think debate is necessary in order for our democratic system to function. In this context, the task of PR is to mirror democratic structures of advocacy. Public relations creates, shapes, and promotes a politics that is embedded in our major institutions, our common practices of mediated debate, and even the way we collectively think about what "the public" is and what it ought to do. It represents various viewpoints, provides information, and solicits opinion. That this information and these viewpoints are plainly unscientific, that these truths are clearly "inconvenient," becomes less important than adhering to the values of democracy.

Public relations is not invested in truth. It is invested in legitimacy. And legitimacy is a relative proposition, or as Mark Suchman suggests, a pragmatic concept.[14] It isn't a quest for truth or facts. It's a way to *use* concepts of truth or facts to persuade others that your view is the best one, in order to gain support in a particular context. This is one reason that the ongoing belief that more and better information about global warming will spur publics to action has not been realized. This belief misrecognizes the role of PR in establishing legitimacy for its representatives through appeals to the public, to information, and to democracy.

PR as a Technology of Legitimacy

In this book, we treat public relations as a technology of legitimacy. The etymology of the word "technology" is relevant: the science, or logic (-*logia*), of skill or craft (*techne*). We examine the logics by which public relations agents developed their craft over the course of the twentieth century. The aim is to show how this craft—the knowledge, tools, and techniques invented and applied to creating relations with determinate publics—became central to the operation of democratic publicity, even as its mechanics were obscured from public view. At issue is the question of how democracy came to include these features rather than work against them. If part of our mission in this book is to move beyond arguments about PR as manipulated publicity, another part aims to avoid replicating the strain of thinking around PR, mainly by PR people themselves, that PR is necessary or important for advancing democratic virtues and values. Rather, our goal is to reveal how PR worked through these values to present its objectives as aligned with the social and political concerns of the time, including the role of publics in a democracy.[15]

Our conception of public relations is therefore ontologically different from what most PR research considers. Our aim is to theorize PR at a systemic level. We show how it established structures of advocacy that legitimized particular rationales for action and strategies for management of information within existing understandings of democratic communication. These structures include the development of political and social institutions.

For instance, PR actors played instrumental roles in the social communication practices of trade associations, helping them to recognize themselves as public advocates for not only their company members but also for an economic and political system that supported their industries. These advocacy structures owed their success to more than their economic and political relevance. Key to their legitimacy was PR innovators' ability to navigate the cultural and moral environment in which they operated. A coalition of companies organized around preventing legislation to limit fuel emissions has to be recognized as fitting into the existing architecture, norms, and standards of political discourse in order to be taken as legitimate. By the same token, calling that coalition the Coalition for Vehicle Choice has a cultural valence that speaks to (and helps produce) values and beliefs of the era. Our approach is therefore both material and cultural: to inquire into both the formation of advocacy structures and how they are wielded, and also into what makes them meaningful in a given time and place.[16]

Considering PR as a technology of legitimacy refers not only to securing legitimacy for one viewpoint over another. It is also about how this business has created a set of social and political conditions in which certain ways of thinking become available to us while others are foreclosed. PR is a process that provides conceptual repertoires, repertoires that have influenced how we define public information and communication around environmental change. Limiting the analysis to manipulation misses out on the specific ways PR embeds itself into our sense-making.

In her stunning book, *Strangers in Their Own Land,* sociologist Arlie Hochschild refers to environmental problems like industrial pollution and ecosystem destruction as a "keyhole issue." Looking through the keyhole of fracking and petrochemical use in Louisiana allows her to develop a cultural understanding of American politics, by showing how charismatic leaders, community associations, and ordinary citizens ascribe meanings to politics that fit their own senses of self and society. At one level, the natural environment is the keyhole issue in this book too. By tracing the transformation of the discourse of environmentalism over the decades of the twentieth and twenty-first centuries, we see to what extent making the environment meaningful to different publics was historically conditioned by the work of professional public relations. But just as the key has no function independently of its keyhole, so do we need to contemplate the lines of thought that make up environmentalism to make sense of the role of public relations in American politics and culture.

PR Counselors as an Epistemic Community

Who are these PR guys engaged in the war of information around climate change? What drives them to do what they do? How have they maintained their operations behind the scenes, and with what impact on public communication? And to what extent have their actions informed, and been informed by, the historical legacy of environmentalism throughout the twentieth century?

To fully account for the evolving strategic nature over the course of the twentieth century requires careful attention to those doing the strategizing. Most writing on public relations considers PR agents as value-neutral information intermediaries who work on behalf of clients to convey their ideas to their chosen publics. In contrast, we consider PR agents to be value-generating actors who create and shape cultural narratives, information standards, and rules of engagement with strategically framed interlocutors.[17]

In this book, we consider PR consultants to be an epistemic community. Defined as self-structured groups sharing professional expertise, beliefs, and common objectives for influencing public policy, epistemic communities claim authority over expert knowledge and seek to embed this legitimacy into their objectives.[18] As knowledge-based networks, epistemic communities also influence meaning-making processes by circulating particular understandings of issues among different publics.[19] Seen in this light, PR is not only about the communication of ideas and information but also about creating the ideas and information standards that shape political contexts.

The ability of public relations actors to present themselves and their work in terms of facilitation and amplification rather than innovation and authoritative direction is a defining characteristic of promotional industries more generally. The "transnational promotional class" is made up of self-styled intermediaries such as lobbyists, consultants, public relations practitioners, and marketers who present their work in terms of brokerage between political figures and their publics. These promotional elites professionalize, mediatize, and manage the process of political communication and policymaking. They do not form a self-consciously composed collective entity but rather operate as a loosely affiliated coalition of actors and institutions dedicated to constructing and managing international and domestic public opinion as well as the conditions in which public attitudes and values are collected.[20]

To understand the effect of this collective is to focus on the ways that this self-conscious intermediary role allows PR actors to carry out their work. In *The Politics of Misinformation*, Murray Edelman argues that publicity operates according to a paradox: political information and actions that are publicized are rarely those that lead to actual social change:

> Political actions, talk, and media reporting focus largely on elections, legislation, and the publicized promises of officials, candidates, and interest groups. All of these institutions emphasize their support for needed change and the reality of change, but none of them makes much difference. By contrast, the activities that do make a substantial difference are largely unpublicized, or redefined as something different from what they are.[21]

A central aim of this book is to bring out these unpublicized activities, by attending to the actors who create and justify them. In order to account for publicity as a technology of power and influence, attention must be paid to the strategies and motivations of those who deliberately avoid the limelight. Examining PR actors' strategies of silence helps us to show how this boundary work allowed them to build up their professional repertoire and gather insights from across sectors.

As market intermediaries, PR counselors occupy a very important liminal position among industries and between industry and government. Like lawyers or accountants, they work across industrial sectors with a wide range of organizational clients. They can build thematic expertise and knowledge that gets solidified into rules and standards, and these get carried across their client base. Unlike lawyers or accountants, however, they are not bound in the same way by the law or the tax code, so they have considerably more flexibility in generating ideas and information for different audiences.

This flexibility is further reflected in the networks of legitimacy in which PR counselors operate. The effectiveness of public relations in the realm of environmental politics is necessarily embedded in a much wider ecosystem of influence: trade associations, industry or science advisory councils, think tanks, research institutes, nongovernmental organizations (NGOs), foundations, chambers of commerce, and organizational boards.[22] While many of these institutions are themselves limited in scope because of their focus on a single trade, industry, style of research, or membership group, public relations practitioners can move freely among them, maintaining multiple affiliations and a broad client base. In the context of environmental issues in the time period covered here, the cohesion of these networks has been to a great extent coordinated by PR actors, who have used their liminality and invisibility to move among network nodes. The trans-industrial and transnational coordination of this network helps to explain the remarkable ideological convergence of corporate public relations across industrial sectors, firms, and national boundaries.

Another explanation lies in the charismatic personalities of some of the more prominent public relations actors. Bernays's larger-than-life personality was paradigmatic of many twentieth-century public relations counselors who innovated in the realm of environmental communication. How such charismatic figures managed to promote themselves and their industry while maintaining the profession's secrecy is part of the story we wish to tell. Indeed, a tension we explore in this book is how the self-aggrandizement of PR actors and PR literature fits within the industry's ongoing attempt to maintain its distance and status as neutral facilitator as opposed to powerful intervener in political affairs.

One outcome of such lofty self-representation is continued slippage among the industry's terms of engagement. Public relations, public affairs, advocacy, lobbying, and the notion of strategic communication more broadly have ambiguous boundaries that often remain unobserved in practice. In this book, these terms are all in play. We use them according to the ways they are deployed in empirical situations. To give a prominent example: the Watergate scandal was a point of inflection for the public relations industry. Increasing public scrutiny in the mid-1970s, as well as congressional reforms distributing power across subcommittees, made old-style centralized lobbying ineffective. For many professional communicators, the solution was to divest one's consultancy of its lobbying function—in some cases, by founding a separate agency to keep the lobbying payments at arm's length from the firm. For others, the solution was to integrate PR and lobbying in new structures of advocacy, such as by forming and sponsoring grassroots constituencies of local citizens who could call on their congressional representatives directly.

Public relations counselors' skill set is therefore deeply contingent and constantly evolving. Professional titles change regularly, as do the toolkits used to represent client needs. In twentieth-century settings of environmental concerns, public relations counselors continually looked for ways to make the environment manageable to their various publics. While in some instances this meant decoupling client activities from environmental issues altogether, in others the chosen method might involve concerted efforts toward transparent communication with communities affected by environmental damage.

"Informating" Environmentalism

Ultimately, making the environment manageable required a concerted focus by PR agents on what the anthropologist Kim Fortun calls the "informating" of environmentalism: producing knowledge about the environment that appeared palatable, tangible, and rational.[23] The specific objective by PR actors was to turn environmental problems into problems of information. In this way the actors could intervene, using their expertise to provide the "right" kinds of information in order to control the outcome. But as Fortun observes, informating environmentalism influences what counts as necessary knowledge. It changes how the environment is conceptualized as a problem and who invests in that problem. It shifts notions of risk, sustainability, and responsibility away from its object and onto different terrains of understanding that are directed away from environmental or climate action.[24]

This shift, we argue, is not a recent development. It has taken place in various forms and by various means over the course of the twentieth century. Taking this historical approach allows us to show the incremental ways by which the environment became, for many American publics, the wrong kind of problem: a problem of information, politics, and publicity instead of a problem of our continued existence. The role of PR is, if not singularly responsible, at least centrally involved in this process. Bringing their work to the forefront of our investigation allows us to build on but move beyond recent work on conservative think tanks, skeptical science, and corporate advocacy in US environmental politics.

Constructing PR as a Reflexive Field

One more word on our approach to public relations in this book. Typical evaluations of PR, especially in the realm of environmental politics, limit the scope to corporations and business. When environmental organizations, public interest groups, or civic entities engage in public relations, as all of them do, the practice is distinguished from that of its corporate counterparts through labeling (e.g., advocacy versus propaganda). This distinction is not simply a matter of perspective. Historically, a great number of PR innovations and techniques in the United States were developed with business and profits in mind.[25] The monopoly companies of the early twentieth century in environmentally compromising industries like rail, steel, and coal faced considerable anxiety among Americans over their size and power. Corporate public relations emerged out of this anxiety, charged with a mission to invest the corporation with a "soul."[26] While the concept of "the environment" as a social and moral problem would not be named until the 1960s, many prewar public relations campaigns focused on mitigating the noxious effects of the corporation in their communities, whether direct ecological effects like pollution and waste management or indirect effects such as employee health and welfare. These problems animated the efforts of Progressive era "muckraking" journalists such as Ida Tarbell and novelists like Upton Sinclair, who directed their ire at the corrupting influence of fossil fuel companies.[27] In the second half of the twentieth century, mounting public awareness of ecological harm caused by extractive and air/water-polluting industries (fossil fuels, chemicals, tobacco, nuclear energy) made corporations into symbols of destruction and targets for reform. Again, public relations counsel played instrumental roles in the reorientation of corporate activities in public and political spheres.[28] As the industry developed its professional associations, journals, and academic programs, these were overwhelmingly focused on the functional and administrative goals of private organizations. Such a lopsided perspective is consistent today.

While it is incontrovertibly true that anti-environmental communication, including the manufacture of doubt and outright climate denialism by contentious actors, has taken up an outsized portion of the communications landscape, to focus exclusively on this communication as the legacy of PR reduces the potential for analytical traction. Despite the clear difference in resources, motives, and information content that attends the practical application of public relations by different groups, it is nevertheless vital to pay attention to non-denialist and non-corporate uses of PR for two important reasons. First, to assume that public relations is the sole province of business is to ignore the vastly important role of information management within government, media, and civic organizations in the conception and communication of public problems. In this book, we draw on a growing literature about the use of public relations and strategic communication by civil society actors in the name of social reform as well as multiple interviews with strategic communicators at national and international environmental nonprofits, activist groups, state and local government departments, intergovernmental organizations, and academic and media institutions to show how various organizational actors consider what it means to communicate about the environment, and how they compete or collaborate in different settings.[29]

A second reason is to overcome the limits of a dichotomous analysis of attitudes toward climate change, which makes corporate and non-corporate participants into antagonistic opponents and maintains the political polarization that has come to characterize this sphere of understanding. For example, scholarly arguments about reflexivity as a necessary precondition for apprehending the human, economic, and technological causes of climate change tend to identify as anti-reflexive those defenders of the current system: primarily conservative and corporate entities.[30] This line of thought retains a barrier between each camp: anti-reflexive orientations lack the progressive, moral, and intelligent vision embodied by the reflexive approach.

In practice, however, there is dimensionality *within* groups in addition to differences between them. For instance, ExxonMobil and Shell Oil Company do not always embrace the same tactics. Lumping them together as "industry" or "corporate" actors misses important insights about how different actors overlap in their interests and collaborate or co-opt each other's maneuvers.[31] Further, distinguishing corporate from non-corporate action is not always obvious. When wealthy families allocate portions of their private fortunes to groups fighting industrial regulation, or political action committees amass individual donations to fund advertising for oil-friendly political candidates, determining whether this constitutes specifically corporate action becomes complex.

In the realm of grassroots mobilization to promote citizen involvement in public policymaking, the picture is even more blurred. While some research has uncovered the "astroturf" nature of citizen groups, pointing to their corporate underwriting or to the professionalization of mobilization strategies via so-called grassroots lobbyists (paid public affairs consultants who incentivize and organize citizen participation using a grassroots repertoire), citizen participation is not uniformly instrumental or manufactured. Researchers such as Tim Wood have shown that the motivations and actions of individual participants in industry front groups are often genuine, civic-minded, and morally inspired.[32] Third-party groups may be top-down, but they are not necessarily corporate.

Another limitation of the reflexivity/anti-reflexivity position is that it does not recognize that various fractions of capital have different vested interests. Consider, for instance, the ways that the Keystone XL Pipeline debate mobilized participants who hold rail interests against participants who hold oil interests. The same is true of members of the anti-capitalist movement: the demands and interests of environmental organizations and social movements vary according to their objectives. A historical and contextual approach allows us to move beyond the dichotomy of good and bad actors that has contributed to the antagonism preventing action on climate change in the current American setting.

The point here is not simply that everyone does PR or that context matters. The point is that PR *itself* has played a non-negligible role in maintaining this dichotomy of good and bad actors around environmentalism. It has long served the interests of public relations actors to develop clear enemies in order to sharpen information and communication practices against them. This enemy construction is both particular ("Bill McKibben") and general ("the Left," "activist," "the public"); it affects both the individual and the category.[33] Such constructions serve to build the category of the other actors in this network—the oil industry, or the average citizen, for instance. Paying attention to PR therefore requires attention not only to differentiated uses of strategic communication by a wide range of actors but also to the ways that the PR industry has developed and maintained actor categories as well as dimensionalities within them.[34]

Methodological Considerations

We are two authors with diverse backgrounds and nationalities. Our intellectual training spans the disciplinary subfields of environmental and cultural sociology and the interdisciplinary fields of media studies, cultural studies, and environmental communication. In this book, we make use of approaches and materials derived from all of these contexts of inquiry, and we also draw on practitioner perspectives in journalism, management studies, and public relations. Our research strategies combine ethnographic fieldwork, interviews, and archival methods to develop a broad cultural and historical context.[35] Chapters 1, 2, and 3 make extensive use of company and trade association archives and government records to develop their arguments. Chapters 4 and 5 draw on professional journals, industry reports, and news

coverage of environmental issues. Chapters 6 and 7 present interview material and on-site observation at events with communication strategists, public
relations professionals, advisors, project managers, and environmental
advocates working within a broad range of organizations. Some readers may
find this blended scholarly lineage a little too promiscuous for their taste. We
preferred to sacrifice the discipline that disciplinary boundaries provide for
the more pressing goal of bringing together multiple perspectives on the singularly complex and intractable problem of environmental degradation and
a changing climate. One thing this book tries to make clear is that reckoning
with the problem of the environment requires dialogue among and participation by all people and all perspectives, even—or perhaps especially—those
that have historically appeared antithetical or antagonistic to the cause.

   For this reason, rather than demonizing actors (e.g., corporations) or categories (e.g., spin), *A Strategic Nature* sets a baseline for concerns that have
consumed actors and framed categories for over a century regarding the role
of human beings within their environment. What Americans have come to
think of as "the natural environment" or "the climate" has been forged at the
intersection of a particular conception of information, communication, and
politics in a particular kind of democracy. Different actors, working with
very different motivations and repertoires of action, have tried to influence
the shape of the natural environment, usually to their own advantage. We try
to render here the various efforts to influence, inform, and manage the concept of the environment within this setting and to examine its impacts on the
social imaginary.

   That said, we do not adopt an objective or uncritical stance toward the
problem. It is important to make clear some of the limitations of the project
and its data at the outset. For one thing, the industry of PR is not only not
neutral in its strategies of legitimation, as we detail throughout the book; it
is also not at all diverse in its constitution; and it is often willfully blind to
its exercise of power. Critical public relations scholars have spilled considerable ink to demonstrate the homogeneity of race and class and the gendered hierarchy of industry practitioners; the lack of recognition of power
dynamics in its theories, models, and practices; and the sanitized charac-
terization of PR audiences, or "stakeholders," which perpetually leaves out
populations, territories, and practices that do not seem to fit within the
frameworks created.[36]

   In offering a historically informed perspective on the double evolution of
publicity and environmentalism, we do not take this monolithic aspect of the

field for granted; indeed, the narrowing of the concept of environmentalism we detail here is a direct reflection of this homogeneity and anti-reflexivity. *A Strategic Nature* aims its critique squarely at the particular nexus of environment, information, and publicness that has given rise to a fairly toothless, isomorphic, and jejune discourse in the democratic public sphere. But we also recognize that the lack of diverse voices in this narrative deserves considerably more attention than we are able to devote space to here.

Structure of the Book

Each chapter in the book attends to a particular historical moment in American life where ideas about publics, information, and the environment came together. These historical moments are organized around periods of political contention, where various groups—corporate, civic, professional—saw the need to transform the rules governing American society. In the early part of the twentieth century, for example, the organization of private interests—industry, railroads, and utilities—spawned fear and alarm among citizens, not least for their polluting ways. In response to the intensive political power and influence of industrial monopolies, individuals organized their own "people's lobby" wielding a "new currency of political influence [that] included procedural mastery, technical expertise, and the ability to mobilize public opinion."[37] Now-famous muckrakers like Ida Tarbell used investigative journalistic means to expose the machinations of major polluters like Standard Oil, further paving the way for a culture of reform. As these actions came into the public eye and contentious collectives emerged to demand change, the need for information to regain control of public narratives became more evident.

Chapter 1, Seeing Like a Publicist, locates the origins of public relations alongside emerging environmental narratives at the beginning of the twentieth century. The United States Forest Service, a federal bureau established during Theodore Roosevelt's presidency, represented a vision of nature as resource for development, at odds with the romantic spirit of wilderness preservationists such as John Muir. Chief Forester Gifford Pinchot developed sophisticated mechanisms and messages to promote his commitment to a distinctly American culture of nature, qualifying and transforming the character of environmental information to the news-reading public in the process. Pinchot developed foundational concepts and practices of public relations that would leave deep grooves in the American experience of environmentalism.

In chapter 2, Bringing the Outside In, we examine the industrial infrastructures within which the burgeoning profession of public relations coalesced: rail, steel, and coal, and the simultaneous development of information infrastructures to situate these industries as paragons of democracy in the American imagination. It was in the struggles over labor rights, workers' rights, employee welfare, and industrial reform that the practice of public relations forged its methods, as scions of power and privilege attempted to manage the external environment of public and political opinion to reduce the friction for the machinations of heavy industry. While the external environment does not directly map onto the natural environment, we see in these struggles the porousness of the boundaries between the inside and the outside of industrial production, allowing industrial leaders to control the outside world in addition to the one within their walls. As later chapters will show, this maneuver laid the groundwork for the idea that specialized knowledge of communities' air, land, and water could come from industrial research. The chapter reviews the efforts of now-infamous PR men Ivy Ledbetter Lee and John W. Hill of the firm Hill & Knowlton to develop principles of "industrial democracy," introducing statistical reasoning, third-party promotion, and internal (employee-oriented) publicity programs as part of an ongoing project of fact-making around the benefits of business for American democracy.

Chapter 3, Environment, Energy, Economy, pursues these ideas into the post–Second World War setting, as industrial PR practitioners in the 1950s and 1960s apprehended the formidable rival of environmental pollution and its discontents. Prior to the war, industry was the leading source of information on air pollution among other problems of "industrial hygiene."[38] By bringing environmental problems inside the firm, companies defined both the problems and the solutions to environmental degradation. In the postwar era, however, with new federal science funding, changing norms of media representation and news coverage, and rising legal battles for companies over wartime reparations, alternative voices began to emerge around environmental issues. Amid the transformation of the nature of evidence in postwar scientific research, coupled with a growing public anxiety over depletion of the commons, public relations counsel set out to balance the scales in their corporate clients' favor. They would find this balance in the notion of energy as its own scarce resource in need of protection. The chapter reviews

the expansion of public relations networks and the adoption of environmentalism as a force to be strategically managed.

The creation of the Environmental Protection Agency (EPA) in December
1970 signaled a new era for environmentalism. The role of the EPA was
quickly labeled "command and control" by the industries who stood to be
most affected by the agency's powers. Over the next decade, the rise of puni-
tive (and sometimes retroactive) legislation to hold liable polluting entities
led contentious industries to fight back. One response was the use of public
relations techniques to foster increased dialogue leading to compromise or
collaboration among oppositional parties. In chapter 4, PR for the Public
Interest, we review the endeavors that allowed industrial interests to promote
their anti-environmental agenda as rational and reasonable. It also allowed
them to advocate against the passage of further legislation. By advancing a
rhetoric of "compromising for the common good," PR actors participated
in both defusing the appearance of adversity in a 1970s and 1980s context
of public concern over environmental damage and in cementing public relations as a legitimate profession with specialized skills of negotiation and
dispute resolution.[39] Throughout the 1950s, '60s, and '70s, as battles over environmental futures intensified between environmental groups and business
associations, PR actors sought to create and manage influence in political
contexts. PR consultants developed single-issue coalitions, public-private
partnerships, green business networks, and other multiple-member groups,
along with multi-pronged media strategies, to advance the idea of plurality.

So, on the one hand, corporate PR counselors succeeded in taking control of environmental issues by framing corporate responses to environmentalism in terms of existing cultural structures in the post-Watergate
era: transparency, public participation, and the public interest. On the other,
they self-consciously applied those same values to their craft, conceptualizing PR as a concerted system of information management rather than an
ad hoc process of persuasion. The same sentiment accompanied their work
on environmental issues. To make the environment more tangible, manageable, and measurable, PR counselors developed benchmarking metrics, reporting techniques, certification schemes, and self-auditing logistics. These
maneuvers allowed PR to further portray environmental politics as information politics.

In chapter 5, Sustainable Communication, the role of PR firms as international knowledge brokers is given its due. The chapter demonstrates the
impact of a network of American public relations firms in spreading "green"

PR across European and Mexican borders during a critical historical period. With the consolidation of the European Union and NAFTA on the horizon, corporate clients in a range of industries (from tobacco to chemicals to oil, coal, and gas) adopted promotional methods that advertised their commitment to environmentalism in an effort to sidestep sweeping regulations. By diffusing its core principles of sustainable *communication* over sustainable environmental behavior, PR networks helped to define environmental communication as a field in its own right, acting as a major cultural producer in the realm of international environmental governance.

In chapter 6, The Climate of Publicity, we examine the media plans, mobilization efforts, and marketing devices that climate advocates use to promote "the planet" to various publics as an object of concern. We begin by asking what it is that PR "knows" about environmental advocacy. While PR appears in the world as a neutral technology of legitimation, this chapter demonstrates the extent to which the practice is culturally determined and how its conception of publics as situational, contingent, and self-interested plays out in its operation. Drawing on interviews with environmental advocates, movement leaders, NGOs, and climate communications teams, we then show how PR, conceptualized by environmentalists as a strategic resource against established systems of power, ultimately reproduces those systems of power, leaving unchanged the substance of response to the "super wicked" problem of climate change.

Chapter 7, "Shared Value": Promoting Climate Change for Data Worlds, begins with a provocation. In the growing movement to deploy big data for big solutions to mitigate global warming, is the data serving the climate cause? Or is the climate a convenient form of promotional capital for the benefit of big data adherents? This chapter reviews the shape of the Data for Climate Action (D4CA) campaign, showing how the campaign's greatest impact is in the realm of publicity. Under the banner of shared value and social good, business, NGO, and political leaders promote data solutions to climate problems, privileging technical and private sector expertise and digital "evidence" of global climate transformations. Despite its datafied package, the chapter reveals the continuity of mechanisms of public relations to generate facts that further reinforce the informational and technical character of environmentalism.

It is not particularly novel to say that the communications or mediation work of organizational actors matters for how we think about the environment. Studies across the academy have looked extensively at how environmental concerns have been fostered, shaped, and influenced by mediated representation in various forms. Dedicated work has been conducted on environmental framing and its consequences; the disciplining discourses of environmental governance; the professionalization of strategic communication and its uses for public opinion and policymaking around environmental issues; the rhetorical and image strategies deployed to promote environmental values and beliefs; and the technologies of environmental knowledge-making, such as modeling, mapping, and monitoring, among many other approaches.[40]

What has not been given its due is the specific role of the public relations industry in making the environment into a matter of concern. The task of this book is to show the historical co-evolution of environmental publics and publicity with the public relations industry and how this co-evolution impacts our contemporary thinking about environmental change.

The environment is a special case of political contestation, because it is not at root a political problem. Showing what role the PR industry has played in turning environmental problems into other kinds of problems—political problems, problems of information, problems of individual attention, in short, into anything but an environmental problem—is the aim of this book. This has meant that generalized expressions of environmental concern, such as mobilization for collective action, ethical commitments to lower consumption and take personal responsibility, and values of pluralist participation and organizational transparency, have been narrowed to fit into advocacy structures that rely on publicity and its subjectivist reorganizations.

If it is true that "publics do not exist apart from the discourse that addresses them," the kinds of environmental publics that PR brings forward are beholden to a limited discourse that is not open-ended, reflexive, or accessible.[41] In Habermas's conception, this is the essence of "manipulative publicity"—a stylized censorship of the free provision of information necessary to a participatory democracy. His ideal of the public sphere, in which individuals come together in public settings to debate, transform, and criticize ideas, is quashed by the presence of large-scale organizations, including the state and corporate power. The reason and criticism necessary to ensure a robust public conscience as a countervailing force to power was suppressed by powerful self-interested groups.

But what if our current model of participatory democracy is constituted by this "manipulative publicity"? As the historian Timothy Mitchell argues in his book, *Carbon Democracy*, "The term 'democracy' can have two kinds of meaning. It can refer to ways of making effective claims for a more just and egalitarian common world. Or it can refer to a mode of governing populations that employs popular consent as a means of *limiting* claims for greater equality and justice by dividing up the common world."[42]

A Strategic Nature builds on that idea by inserting the determinate role of public relations in making this relationship between carbon and democracy legible and palatable to modern publics. More to the point, it is about the role of public relations in creating the publics necessary to accept this relationship. If democracy is characterized by Mitchell's second definition, then, he argues, "the problem of democracy becomes a question of how to manufacture a new model of the citizen."[43] We see public relations as instrumental to this process.

1

Seeing Like a Publicist

How the Environment Became an Issue

> For ultimately all consequences which enter human life depend
> upon physical conditions; they can be understood and mastered
> only as the latter are taken into account. One would think, then, that
> any state of affairs which tends to render the things of the environ-
> ment unknown and incommunicable by human beings in terms of
> their own activities and sufferings would be deplored as a disaster.
> —John Dewey, *The Public and Its Problems*

No history of American public relations is possible without acknowledgment
of the importance of the Progressive era. It was during this time in the early
twentieth century that the idea of the public as a check on power, a source of
truth, and a mainstay of democratic life was formed. The Progressive reliance
on the public as the nation's conscience was intimately connected to the role
of publicity as an instrument of truth in the service of reform. To publicize
was to disclose, reveal, and educate, for it was "through the laying out of material facts and the publishing of information [that] the public would become
activated" to bring about democratic change.[1]

Publicity in itself was important, but so was the form in which this publicity was disseminated. In order for information to appear as a public issue
worthy of attention in this era, it had to be made visible in a particular way.
It had to appeal to an audience that was newly massified but not yet self-
consciously national or integrated. It had to appear to push back on behalf of
citizens against abuses of power perpetuated by monopoly interests. It had to
show, as the progressive reformist and journalist H. D. Lloyd wrote in 1881,
"the points where we fail, as between man and man, employer and employed,
the public and the corporation, the state and the citizen, to maintain the
equities of 'government'—and employment—'of the people, by the people,
for the people.'"[2]

A Strategic Nature. Melissa Aronczyk and Maria I. Espinoza, Oxford University Press. © Oxford University Press 2022.
DOI: 10.1093/oso/9780190055349.003.0002

This form was the popular press. At the turn of the twentieth century, the news media emerged from its primarily "aesthetic" role to become the seat of social responsibility, transparency, and ultimate commitment to the project of democracy.[3] News was at once technology and cultural form: both cause and effect of a class of information considered incontrovertible fact. If publicity was the "great moral disinfectant," as Lloyd put it, the news media was its righteous reflection.[4]

For an idea to be known and communicable by human beings as part of their own social responsibility, to paraphrase Dewey, it had to be made into a public issue. And to be made into an issue, it had to become a matter of public interest over private gain, of communal knowledge over hidden intentions, and of popular sovereignty over political or commercial machination. In other words, it had to be made into news.

This chapter is about how the environment became just such an issue and what happened when it did. The naturalist John Muir and the forester Gifford Pinchot are frequently hailed as instigators of a twentieth-century national consciousness around the need for protection of the natural environment. Both advocated powerfully and persuasively for the conservation of land and its benefits to Americans. Both made use of extensive publicity via multiple forms, including the news media, to identify the environment as a modern public problem. What would become clear, over the decades devoted to their cause, was how publicity could be invoked to promote collective participation and a sense of shared obligation or to reinforce existing structures of power and expert authority. Seen as a matter of public concern whose resolution is subject to popular decision, the power of information lies in its wide distribution and not in the control over how that information is received. But seen as a matter of expert administration, where only certain people are deemed qualified to exercise judgment, information becomes a resource for achieving determinate goals. The natural world as an idea has historically raised a parallel question of who ought to be in charge of it. And in the attempt to determine whether this world was a matter of public concern or private governance, new strategies of reason and regulation would emerge.

The Wrong Publicity: "Spiritual Lobbying" for the Forests

In the late nineteenth and early twentieth centuries, the environment was not an issue. This is not to say there were no historical precedents for environmental thought. From the time of the first Puritan settlements in the United States in 1620, and for the next 250 years, colonizers saw their "manifest destiny" in terms of control and mastery over nature.[5] In the mid-1700s, cultural movements of primitivism and the romantic sublime waxed poetic about nature with a capital N. In the nineteenth century, myths of the frontier and freedom of the land were painted as characteristics of American identity, and increasing development of the land by timber, mining, and rail interests gave additional meaning to nature as a source of supply. But considered as a set of unified, publicly motivated concerns over industrial pollution, commercial exploitation, and resource extraction—what Lawrence Buell calls a "toxic discourse"—there was no environment.[6]

The strength of each of these early environmental mythologies relied to a large extent on imagining them in terms of absence. To invoke nature was to desire to preserve it amid the threat of its disappearance. An 1893 essay by the historian Frederick Jackson Turner is paradigmatic: The frontier land, "the meeting point between savagery and civilization" and the source of "a composite nationality for the American people," was disrupted by the removal of "frontier" as a category of place in the 1890 census. For Turner, this was at once the end of a major historic movement and a recognition of the central role of the natural environment in American identity.[7]

The trouble with many of these origin myths is their inability to imagine an environment as existing in harmony with human activity. Whether viewed as something to be controlled and overcome in the name of civilization or as a pristine wilderness and haven away from civilization, nature was separated from humans. This is what the environmental historian William Cronon calls "the trouble with wilderness": a paradigm in which wilderness is foundational to American identity and yet divorced from "the material world of physical nature," the social problems of environmental health, and the historical realities of centuries-long manipulation of the natural world. In attempting to preserve these ideas, we are "getting back to the wrong nature," one in which "too many corners of the earth become less than natural and too many other people become less than human."[8]

The legacy of John Muir is sometimes seen as part of this "wrong nature."[9] And yet, this legacy was instrumental in raising awareness of the environment and the threat of its destruction. Anointed as "the father of the environmental movement," Muir was a radical, religious, and romantic lover of wilderness. He was also a prolific and lyrical writer; his books and essays on nature and its wonders were immensely popular in his day. "I care to live only to entice people to look at Nature's loveliness," he wrote to a friend from Yosemite Valley in 1874.[10] He was, in the words of his close friend and colleague Robert Underwood Johnson, "a pioneer of Nature but also a pioneer of Truth," fueled by the depth of his conviction.[11]

Muir made preserving the natural environment his life's work. Committed to national salvation by way of wilderness, Muir advocated tirelessly to protect American land from exploitation. He was highly successful in rallying sympathetic allies to his cause. In addition to the support of the Sierra Club, which he founded and presided over until his death in 1914, Muir could draw on the support of powerful figures in government, media, and the wealthy elite. He even had backing from industrial interests, who saw nature preservation as an opportunity to promote tourism.

Whether or not Muir's legacy illustrates "the wrong nature," his advocacy for nature protection is a clear harbinger of the wrong publicity. He was unsuccessful in his advocacy to maintain the Hetch Hetchy Valley as a national park, one of the defining episodes in American environmental history. For to turn the environment into a public issue, one has to make the environment into an object of politics. And Muir's public relations were not up to the task.

***

John Muir was born in Scotland in 1838 to a deeply religious family. In search of a stricter set of religious teachings than those offered by the Church of Scotland, the family immigrated to the United States in 1849. Inventive, intelligent, consumed by both disciplined instruction and a love of the natural world, he became proficient during his lifetime at botany, geology, chemistry, and glaciology, and he was called by turns naturalist, explorer, philosopher, and transcendentalist. It is an oversimplification to qualify Muir's vision of a wilderness belonging to the people as "preservationism," especially when placed in opposition to a notion of "conservationism" as the managed use of nature as human resource. Themselves products of deliberate publicity, these two positions nevertheless often appear in the current American imagination as divergent paths in the wood, with the one followed and the other not taken.

Historians have called Muir an expert "publicizer" of wilderness; his colleague Robert Underwood Johnson referred to him as a skilled "propagandist."[12] But it was really Johnson himself who was the master publicizer and propagandist for Muir's views. In May 1873, a twenty-year-old Johnson entered the offices of *Scribner's Monthly* magazine to inquire about a position in the editorial office (a position more or less guaranteed by his family connections). For the next forty years Johnson would serve on the staff of the magazine, becoming associate editor when it was renamed the *Century Magazine* in 1881. An illustrated news and current affairs magazine, the *Century* was aimed at a middle-class readership, shaping the opinions of around 250,000 monthly readers, at its peak, on progressive causes popular in its day, such as women's suffrage and civil reform. Along with *Harper's*, the *Atlantic*, and the daily newspapers, the *Century* contributed in no small way to the making of a national American public. As an advertisement promoting subscriptions to the magazine claimed, "The *Century* magazine is doing more than any other private agency to teach the American people the true meaning of the words Nation and Democracy."

As the *Century*'s editor, Johnson saw fit to use the magazine to promote strong views on contentious issues. He also made extensive use of his social and political allies to push hard in Washington for those views. Indeed, the magazine was not merely a source of news but a promotional device for Johnson's extensive lobbying. It published his and others' writing on matters he deemed politically and socially relevant, such as international copyright, the abolition of tariffs on art, and forest conservation. The editorial offices of the *Century* often hosted meetings of committees created to advocate for those purposes—committees he himself had sometimes formed.[13] Johnson referred to these activities as "spiritual lobbying" insofar as they formed "measures of American honor or well-being."[14]

Johnson and Muir met in a California hotel in 1889. A few weeks later, they embarked on a camping trip to the Yosemite Valley and the Sierra. It was here that the idea for Yosemite National Park was born. As Johnson recounts it, on hearing Muir complain about the "hoofed locusts" (sheep) whose grazing had eroded the mountain vegetation and affected irrigation, he proposed that the valley be protected as a national park along the lines of the Yellowstone. He asked Muir to write two articles for the *Century*: one to vaunt the Yosemite's features to a general public, and the other to elaborate a formal proposal for the park.[15] Johnson would take the proofs, along with illustrations of the region, to Congress to advocate for the park.[16]

On 1 October 1890, a bill was passed establishing the Yosemite National Park. With this bill, and with the public attention it had generated, Johnson and Muir now had the beginnings of an information and influence campaign to bring more of California's land under federal auspices. Johnson pursued the campaign with vigor. He enjoined the established author and landscape architect Frederick Law Olmsted to support the cause in the *New York Evening Post*. He encouraged Muir to create a "Yosemite and Yellowstone defense association" that would "enlist the support of the people and the government in preserving the forests and other features of the Sierra Nevada mountains."[17] This defense association would be founded as the Sierra Club, with Muir as its president from its inception until his death in 1914 (with Johnson as honorary vice-president).[18]

In 1895, Johnson published a series of short opinion pieces under the title, "A Plan to Save the Forests," in the *Century*. The plan was for "a thorough, scientific, and permanent system of forest management in this country."[19] The opinions were furnished by Muir; Olmsted; Edward A. Bowers, assistant commissioner of public lands; B. E. Fernow, chief of the Division of Forestry in the federal Department of Agriculture; and a number of other well-placed supporters, among them Theodore Roosevelt, then with the US Civil Service Commission.[20] Not surprisingly, all of them were in support of a particular vision: the army should take charge of guarding the forests; and the academy of West Point should initiate a training program in forestry. The magazine's editorial urged "the appointment by the President of the U.S. of a commission composed of men of sufficient reputation to make their recommendations heeded, whose business it shall be to study the whole question of forest preservation, and to report fully on it to Congress."[21]

It was through the ensuing commission that Muir and Pinchot were introduced. Pinchot was secretary of the commission and Muir an unofficial consulting member. At the end of their field investigations, the commission members disagreed on what to prioritize in the report: more preservation, along the lines of the 1891 Forest Reserve Act (which Johnson had helped lobby for, using sketches of the King's River canyon provided by Muir)? Or a more "practical," managed approach, with room for economic development of the forests? The former view initially prevailed, with President Cleveland setting aside more than 21 million acres of forest land in his final days in office; but the resultant hue and cry from the western states and from commercial interests pushed the new administration to approve the latter idea.[22] The regulatory power of the resultant Forest Management Act of 1897 was limited, as it was clearly designed to appease western legislators as well as curtail restrictions for lumbering, grazing, and mining.[23]

In the interim, Muir had written articles for the *Atlantic* and *Harper's* magazines in which he equivocated somewhat, hoping a unified view of forest protection would be more effective against industrial concerns.[24]

When the Forest Management Act was passed, however, he took off the white gloves. In "The Wild Parks and Forest Reservations of the West," published in 1898, he leveled his criticism at both political and commercial interests:

> This Sierra Reserve, proclaimed by the President of the United States in September, 1893, is worth the most thoughtful care of the government for its own sake, without considering its value as the fountain of the rivers on which the fertility of the great San Joaquin Valley depends. Yet it gets no care at all. In the fog of tariff, silver, and annexation politics it is left wholly unguarded, though the management of the adjacent national parks by a few soldiers shows how well and how easily it can be preserved. In the meantime, lumbermen are allowed to spoil it at their will, and sheep in uncountable ravenous hordes to trample it and devour every green leaf within reach; while the shepherds, like destroying angels, set innumerable fires, which burn not only the undergrowth of seedlings on which the permanence of the forest depends, but countless thousands of the venerable giants. If every citizen could take one walk through this reserve, there would be no more trouble about its care; for only in darkness does vandalism flourish.[25]

Muir also saw fit to do some personal advocacy, conducting "campfire diplomacy" on trips to the Yosemite with Theodore Roosevelt. This had some benefit: President Roosevelt supported Muir's idea to incorporate Yosemite Valley into the existing national park; and at Muir's urging, Roosevelt in 1908 designated the Grand Canyon a national monument.[26] Roosevelt's love of nature was well known. He had founded the Boone & Crockett Club in 1887 as a hunting and fishing group and regularly spent time out of doors. Roosevelt's biographer, Edmund Morris, recounted one of Roosevelt's Yosemite escapades with Muir, spending days and nights in awe of their wild surroundings. During the trip, Muir talked nonstop about the need to preserve the environment in which they reveled. Although it appears that Roosevelt "would have preferred to hear less of Muir and more of the hermit thrushes," he was also taken with Muir's "pure form of preservation" against the utilitarian "greatest good for the greatest number" ("conservation") perspective embodied by his friend and chief forester in the federal government, Gifford Pinchot.[27]

It was the argument of the "greatest good," however, that would come to dominate, and ultimately define, the public interest. Though it was Muir who captured Roosevelt's imagination, it was Pinchot who would win

Roosevelt's favor in terms of national policy for the forests. "In matters of forestry," he is quoted as saying, "I have put my conscience in the keeping of Gifford Pinchot."[28] Pinchot's vision for forests was more practical, more scientifically verifiable, and above all, more legible for a political conception of nature. But the real contest, as the next section of the chapter makes clear, was for Pinchot to make this vision more relevant to the broad vision of all Americans as a concerned public and to present it as the more qualified expression of the public interest. It is in Pinchot's clever management of public sentiment that we perceive the emergent politics of a new environmental awareness.

Seeing like a Publicist: State Forestry and the Discipline of Public Relations

It is not for nothing that Lippmann identifies Pinchot as the archetypal expert in his first book, *A Preface to Politics*:

> The statesman acts in part as an intermediary between the experts and his constituency. He makes social movements conscious of themselves, expresses their needs, gathers their power and then thrusts them behind the inventor and the technician in the task of actual achievement. What Roosevelt did in the conservation movement was typical of the statesman's work. He recognized the need of attention to natural resources, made it public, crystallized its force and delegated the technical accomplishment to Pinchot and his subordinates.[29]

A Preface to Politics was published in 1913 at the height of the controversy over the Hetch Hetchy Valley, a controversy called "the spiritual watershed of American conservation history."[30] The damming of Hetch Hetchy to create a water reservoir for California cities symbolizes Pinchot's triumph: nature as utilitarian resource over nature as protected wilderness. As Roosevelt's "conscience" in matters of forestry, Pinchot embodied the specialized technical skills, "the ingenuity to devise and plan," that Lippmann believed was required to turn ideas into practical effects.

Pinchot was indeed an expert in the management of nature. Throughout his long career in private and public forestry, Pinchot transformed the state approach to the conservation of forests and related natural resources. The methods Pinchot advocated are embedded in systems of not only govern-ment but also education and commercial resource use.

Pinchot was also an expert in the management of publics. From the begin-ning of his life as a professional forester, Pinchot engaged in constant pro-motion of his work. His calculus was born partly of conviction but also of a deft awareness of the value of public support for his vision of utilitarian for-estry, and he cultivated it by a dizzying array of means. As he wrote to R. C. Melward in 1903:

> Nothing permanent can be accomplished in this country unless it is backed by sound public sentiment. The greater part of our work, therefore, has consisted in arousing a general interest in practical forestry throughout the country and in gradually changing public sentiment toward a more conser-vative treatment of forest lands.[31]

Making forestry legible meant making it visible in a particular way. It meant "seeing like a state": projecting nature as a project of legibility and simplifi-cation, which can be ordered through a utilitarian, abstract logic.[32] To the extent that Pinchot saw protecting nature as a moral obligation, this was to be established by means of having the facts. It was information, not ethics, that would create the contours of the knowledgeable public.[33]

Muir was the "pioneer," the "prophet."[34] And he was a keen propagandist in his own right. But it was Pinchot who figured out how best to represent the environment to its constituencies. It was through his work that we see the beginning of a mutually constitutive evolution: that of forestry and its "sci-entific" principles of management, which enabled nature to be understood as natural resource; and that of publicity as an institutionalized, rational, and coherent endeavor.

***

Pinchot is known as "the first professional trained American forester."[35] After undergraduate studies at Yale, Pinchot was sent to Europe to learn the methods of scientific forestry. The curriculum had been invented in Germany in the second half of the eighteenth century. By the end of the nineteenth century, it had spread throughout the continent and become "hegemonic."[36]

Pinchot's first opportunity to apply these newly learned methods came in 1892, when he was hired by George W. Vanderbilt to manage the Biltmore Forest on his estate in western North Carolina. Pinchot was advised by

Frederick Law Olmsted to demonstrate how wild nature could be cultivated to look natural. "At Biltmore, [Olmsted] and his protégé Gifford Pinchot would advance an American culture *of* nature."[37]

Promoting the results of his work was part of Pinchot's contract: he prepared an exhibition of his forestry methods for the World's Columbian Exposition in Chicago the following year.[38] The exhibition featured photographs and maps of the Biltmore Forest as well as European forests to be used to model future plans. He also prepared a pamphlet detailing physical features of the forest and specific costs involved in his work to date, of which 10,000 copies were circulated.[39] Reviewing the pamphlet, the popular magazine *Garden and Forest* exclaimed that it "must be considered a most important step in the progress of American civilization, as it records the results of the first attempt that has been made on a large scale in America to manage a piece of forest property on the scientific principles which prevail in France, Germany and other European countries."[40]

Pinchot desired additional means to promote his technical methods. In 1893 he opened an office in New York City and hung out a shingle: "Consulting Forester." Throughout the decade he gave public presentations and advised private landowners on their forests. He also had the opportunity, in 1894, to meet Theodore Roosevelt (while a member of the US Civil Service Commission) and impressed him favorably with his projects.[41]

His first foray into the power of public opinion came in early 1897, when the outgoing US president Grover Cleveland set aside 21 million acres of forest reserves. Amid the outcry in the West, with newspapers objecting to the sudden halt to settlements and western development, Pinchot was sent as "special forest agent" to evaluate the situation.[42] Pinchot leaned on connections with editors and with former Yale classmates at western newspapers to secure copy favorable to the reserves, including the text of an interview he conducted with a writer at one of the papers.[43]

The following year, newly appointed chief forester in the federal government, Pinchot put his developing understanding of the direction of public opinion to work. His first task was to justify the existence of his new department home. Considerable opposition by Congress to both the practicability of scientific forestry and the potential of the government to create policy around the forests had led to calls for dissolving the division. And since the national forest reserves were under the jurisdiction of the Department of the Interior, Pinchot was effectively "a federal forester without forests."[44] To rectify the situation, or, as his biographer put it, "spread the gospel of scientific

forest management," Pinchot adopted a multi-pronged strategy.[45] He quin-
tupled the mailing list for the Division of Forestry, bringing the number of
recipients of forestry information up to 6,000 (including 2,000 newspapers).
He increased the output of publications about forestry from the division and
raised the printing order from 58,000 to 92,500 copies.[46]

Pinchot recognized the benefit of accumulating allies. A particularly
powerful group to bring on his side were private owners of timberland,
with whom he already had some connection and whose managed forests
could then serve as calling cards for his methods. He prepared a circular offering the methods of the Division of Forestry to farmers, lumbermen, and
others who might benefit. Over the next ten years, more than 900 formal
applications were made for the management assistance of the federal government, including among others William G. Rockefeller, E. H. Harriman, the
Great Northern Paper Company, and the Weyerhaeuser timber company.[47]

More influential still than private forestry owners or newsletters for the
making of public opinion was establishing good relations between the division and the news media. The key, as archivist Harold T. Pinkett writes, lay in
convincing the newspapers *that forestry was news.*[48] To present information about forestry within the genre of news was in essence to turn Pinchot's
vision of forestry into fact. Unlike government publications or even magazine editorials, which could seem to a Progressive era public like a statement
of self-interest, making forestry into news was to make it into a matter of
public interest.

Turning information into fact is a process the cultural historian Mary
Poovey calls "factualization." The power of factualization lies partly in the
genre considered most truthful or accurate in a given context. In this era, as
we have seen, "the golden age of journalism" had given rise to a transformed
understanding of the genre of news as transparent and publicly necessary
fact. The effect of factualization is twofold. It creates a certain understanding
of what kind of knowledge is considered legitimate, and it elevates the
knower to the realm of expert.[49]

If factualization relies for its authority partly on the making of information
into fact, it also requires the transformation of other kinds of information
into fiction. In other words, to maintain the legitimacy of some kinds of information, competing versions must be shown to be less legitimate. In this
context, the contest for authority over what kind of nature was most legitimate resulted in a contest over what kind of publicity was most legitimate.
There was the version of publicity synonymous with public reform, "evoking

liberal notions of public enlightenment, press freedom and political accountability," and there was the version of publicity as false coin of self-interested
exchange.[50] This was not only a matter of format and genre; it was also about
whether the publicity appeared to be promoting an ethical truth or a factual one.

The tension between these two forms of publicity became apparent
in 1907–1908, during a congressional inquiry into the activities of the
Forest Service. The sheer volume of information pouring out of the Forest
Service and the apparent use of a "press bureau" from within the division,
an unheard-of use of government resources at the time, elicited concerns
that the chief forester was conducting unseemly practices within the federal
government. On paper, the concerns revolved around costs for publication
and degree of training of personnel; but at root, the inquiry dovetailed into
whether Pinchot was using his political power to promote himself and his
allies first and foremost. It was a fair question. The trade magazine *Irrigation
Age* had called Pinchot "one of the best advertisers of himself and his work
in the United States" and declared that he had Lydia Pinkham, the notorious nineteenth-century marketer of homemade health tonics, "beaten
to a shade."[51] A few years later, an article in *McClure's* magazine titled
"Manufacturing Public Opinion" would call Pinchot "a master and promoter
of political publicity" second only to Roosevelt himself.

Pinchot's appearance before Congress to refute these charges illuminates
the logic of his publicity strategy. The information prepared by the Forest
Service is of the utmost importance in public education, Pinchot countered,
because "the great mass of the American people do not yet understand how
to make the best use of the forest." Limiting information to government
publications was both cost prohibitive and overly technical for a general audience. It was by preparing material for use by news editors, Pinchot claimed,
that the Forestry Service achieved its mission:

> It is not a question of discovering facts and making them known to
> specialists, but of working into the everyday thought and everyday prac-
> tice of great masses of men what the Forester already knows. It is nec-
> essary to convert scientific information into common knowledge. This
> means that not tens or hundreds of thousands, but millions of citizens
> need to be reached. The periodical press of the country affords the best
> means of accomplishing this, since everyone who reads at all reads
> newspapers.[52]

Pinchot also defended his hiring of staff with more newspaper experience than forestry training, arguing that it was necessary to demonstrate the value of the Forestry Division's work to as wide a public as possible. "Above all, the relation between the public welfare and the perpetuation of the forests, the loss of which would mean an impairment of the nation's wealth, will be illumined whenever possible." By presenting his publicity as being in the service of the public welfare and by using the genre of news to frame the publicity as fact, Pinchot succeeded in removing his activities from the taint of impropriety and self-interest, locating them instead as democratic, progressive, and altruistic gestures. This was the "right" kind of publicity.[53]

The outcome of the congressional inquiry—that no federal monies should "be paid or used for the purpose of paying for in whole or in part the preparation or publication of any newspaper or magazine articles"—ultimately benefited Pinchot and his division, since his press bureau did not pay for news coverage but rather gained "free" publicity by allowing its press releases to be picked up and used—or sometimes reprinted wholesale—by the papers.[54] More to the point, Pinchot's press offensive succeeded in establishing entirely new practices of government information and circulation to create and manage informed publics. By bureaucratizing publicity, Pinchot created a systematic, efficient machine to "informate," regularize, and authorize the management of nature. Journalism scholar Stephen Ponder calls Pinchot a "press agent for forestry," arguing that his determined use of government resources to promote his views of conservation in forms "acceptable as news" gave Pinchot license "to dominate discussion of natural resources management at the beginning of the twentieth century and to influence those discussions down to the present."[55] Theodore Roosevelt himself, in his 1913 autobiography, wrote of the Bureau of Forestry, "It is doubtful whether there has ever been elsewhere under the Government such effective publicity—publicity purely in the interest of the people—at so low a cost."[56]

Allies in Environmental Publicity

Associating Pinchot's methods with public reform and enlightenment went beyond the Forest Service's relationships with news editors. The service developed curricula and other educational initiatives to teach principles of forestry in schools as early as kindergarten. A number of the schools used as textbooks Forest Service publications, including Pinchot's own *Primer of*

Forestry, of which more than a million copies were eventually circulated by the federal government.[57]

Pinchot indeed believed strongly that forestry was "something that must be taught."[58] In 1900, the Pinchot family donated a large sum to Yale University to establish the first professional school of forestry in the United States. The donation came with a number of behind-the-scenes ambitions: first, to promote a properly American school of forestry education that would rival prior efforts (at Cornell University by Fernow, of the Department of Agriculture, and at George Vanderbilt's Biltmore Estate by forester Carl Schenck) as well as other proposals, such as the West Point proposal made by Robert Underwood Johnson and Muir in 1895; second, to further professionalize and legitimate the practice of scientific forestry; and third, to create a network of forestry experts to assist Pinchot in his ongoing mission.[59]

One example of these network ties can be seen in the career trajectory of Henry S. Graves.[60] While still a "consulting forester," Pinchot enlisted the assistance of his Yale undergraduate classmate Graves. This assistance initially consisted of fieldwork and the preparation of technical reports.[61] But in 1895, Pinchot paid for Graves to obtain his graduate training in forestry in Europe and hired him on his return in the Bureau of Forestry.[62] When the School of Forestry was founded at Yale, Pinchot installed Graves as the school's first dean.[63] And in 1910, when Pinchot left the Forest Service, it was Graves who would take up the reins, becoming the nation's second American-born chief forester. For years to come, Graves would consult Pinchot to help him "protect the Forest Service from White House influence or congressional machinations."[64]

There were many more: Herbert A. Smith, another Yale classmate, joined the Division of Forestry in 1901 and became publicity director when the press office was created in 1905.[65] Smith prepared drafts of Pinchot's annual reports to the president and the secretary of agriculture and helped prepare some of the president's speeches on matters of conservation.[66] George P. Woodruff and Philip P. Wells, also Yale friends, worked as legal counsel in the Forest Service and for the National Conservation Association, a Washington, DC-based lobby group Pinchot founded in 1909 to push for "effective conservationist legislation."[67]

Again, Pinchot's mastery of publicity was apparent. By creating and institutionalizing a network of allies, Pinchot could draw on these associations to further promote his views. For instance, in addition to monitoring pending legislation on matters of conservation, the National Conservation Association (NCA) distributed regular press releases detailing Pinchot's opinions on the quality of the legislation.[68]

Technologies of Legitimacy in the Forest Service

Pinchot's talents as publicist extended beyond his use of the news. The Forest Service pioneered the use of methods of information management that would later be adopted throughout the federal government. In the interest of greater administrative efficiency, Pinchot adopted a series of recent techno-logical inventions to materialize "an organizational memory" for forestry.[69] Such "systematic management," in the organizational theorist JoAnne Yates's terms, involved extensive recordkeeping via the newly invented vertical filing system, the classification of correspondence by subject, and a system to segregate and dispose of accumulated records "that could very well be destroyed without danger of embarrassment to the Service."[70] This, too, was part of seeing like a state.

To boost recordkeeping potential, Pinchot's press bureau tracked and monitored Forest Service material in circulation. His press bulletins in-cluded tear sheets and were sent out to editors accompanied by postage-paid return envelopes. He also monitored the news via a clipping service.[71] He made extensive use of another "technological marvel": a mailing label machine, which accelerated the printing of addresses for recipients of the Forest Service's reports and bulletins.[72] Pinchot adopted a decentralized administrative structure for his staff of foresters. Foresters had offices throughout the western United States, and Pinchot "deposited forest receipts in the region's banks" to promote greater local acceptance of his forestry practices.[73]

After his dismissal from the Forest Service, Pinchot found ways to main-tain his activities on the "right" side of publicity: his book, *The Fight for Conservation* (1910), was published to "translate[e] these close-quarter struggles with legislators and lobbyists into popular language for a wider audience."[74]

In striving to maintain his advocacy on the "right" side of publicity, Pinchot sought also to accentuate the distinction between "good" and "bad" nature practices. This approach was most dramatically visible in the infa-mous controversy over the use of the Hetch-Hetchy Valley.

The Establishing Act of Twentieth-Century American Environmentalism

Nestled in California's Sierra Nevada mountains, in the northwest corner of the Yosemite National Park, the Hetch-Hetchy Valley was named by the Miwok tribe for the seeds of a grass that grew in the glacier-carved valley. When Muir and Underwood helped establish the Park, the Hetch-Hetchy Valley was to be protected "in perpetuity." The story of how Hetch-Hetchy's waters were redirected into city reservoirs has been analyzed in a number of ways. Some have called it a "national awakening"; others "the single most famous episode in American conservation history."[75] Most have called it a battle: a battle over two staunchly held ideas of nature, encapsulated by the personages of Muir and Pinchot. Indeed, the origin narrative of environmentalism is frequently told through the battle of Hetch-Hetchy, as the triumph of Pinchot's "conservationism" over Muir's "preservationism."

But the real story seems to rely on a battle over something else. That something else is the use of publicity as a technology of legitimacy, advancing one version as more pragmatic, realistic, or feasible than another—and suppressing additional versions, such as that of the land's original inhabitants, in the process. Modern environmentalism is a problem of our continued existence, not a problem of publicity. But in presenting the story of Hetch-Hetchy as a choice over two competing visions of nature, it was made to appear that way. As we have seen, publicity was of utmost importance in the early decades of the twentieth century; but there was a "right" and a "wrong" kind. The right kind was that which was most clearly located in the expert provision of information to generate truths favorable to the exercise of democracy. The determination of whether Hetch-Hetchy should be protected as national park or put in service of supplying water to Californian residents was therefore construed as an act of the people's will based on their response to these truths; and the problem became one of relating to "the public" in a particular way.

Here, the story of Hetch-Hetchy serves two purposes. First, it reveals an emerging professional approach to public relations and its legitimating power in elevating certain kinds of knowledge and expert knowers: those who could create the public interest. Second, Hetch-Hetchy forms the "establishing act" of environmentalism in the modern sense.[76] By the groundswell of attention to Hetch-Hetchy across the United States and by the repertoires of contention mobilized around it, the environment became an issue. And the version of environmentalism that won out set a path for the direction of public and political action around environmental concern for the rest of the twentieth century.

***

The beginnings of the battle over Hetch-Hetchy can be seen in the DeVries Act of 1901, which authorized the secretary of the interior to use rights of way through public lands, including the Yosemite National Park, for electrical and water power infrastructure, provided these uses were "not incompatible with the public interest."[77]

The Hetch-Hetchy Valley had previously been identified by engineers as a possible reservoir site to supply the perpetually dry and sandy California cities. But it was only after a massive earthquake and fire in San Francisco that the city's need for water made national headlines. In 1908, Secretary of the Interior James R. Garfield—a close personal friend of Pinchot's— granted San Francisco the right to dam Hetch-Hetchy and create a water supply.[78]

Muir and his sympathizers—representatives of the Sierra Club, the American Civic Association, the American Historic and Scenic Preservation Society, and the Appalachian Mountain Club of Boston—went before the Public Lands Committee of the House of Representatives, hoping for "a vigorous defense of the people's rights." They made their case along three lines: one, the 1901 rights of way act was not meant to divert large parts of the park from public use; two, the beauty of the park would be ruined and its trails and camping grounds blocked (arguing that the dam would result in "the exclusion of the traveling public and a large army of Summer campers who come there from stifling and dusty lowlands"); and three, other sources of water were available—and had even been proposed to San Francisco—but had not been investigated.[79] On this last point, the Muir contingent implied that special interests may have been at work.

The claim that dam proponents were made up of "special interests" was key to the Muir camp's strategy. The taint of "special interests" was especially strong in this era.[80] Robert Underwood Johnson contributed heartily to this line of attack, painting the proponents of the dam as deep-pocketed Washington insiders who appealed to the administration by turning the decision over Hetch-Hetchy into a question of party affiliation.[81] Johnson and Muir also appealed to an ongoing anxiety of Americans in this era about their waning spiritual commitments. They prepared a number of editorials and pamphlets condemning the materialist tendencies of Americans who valued business and money over more spiritual concerns.

Each side attempted to demonstrate that it was in possession of the facts while the other side engaged in fictions. At times this was framed as a matter of numerical strength, appealing to population size, for instance; at others, possession of the facts depended on physical presence, namely, seeing things with one's own eyes; and in other cases still, the bearer of the facts was the one who was most practical.

These various legitimating and delegitimating tactics appear in a series of opinion pieces about Hetch-Hetchy's fate between Johnson and then-mayor of San Francisco James Phelan in the pages of *Outlook* magazine in 1909. In a sharply worded missive titled "Dismembering Your National Park," Johnson opined:

> It is certain that a rising tide of protest is pouring in upon Congressmen from all quarters against this wanton sacrifice of the public interest. People are asking . . . why the principle of "the greatest good to the greatest number" should merely measure San Francisco's population against actual visitors to the Hetch-Hetchy and not against the whole people.

Two weeks later, in "Why Congress Should Pass the Hetch-Hetchy Bill," Phelan countered that Johnson "speaks of the Valley only by hearsay," having never actually visited it. Moreover, Hetch-Hetchy is only "accessible over difficult trails about three months during the year, and few ever visit it. . . . [T]he highest use of water is the domestic use, and the eight hundred thousand people living in San Francisco and on the opposite shore of the Bay are certainly . . . entitled to the consideration of the country."[82]

Hearsay! Johnson replied, outraged. Johnson had "the testimony of photographs, of which I have twenty, and that of many visitors . . . all of whom have camped in the Valley." He added, "But if I am to be put out of court because I have not seen this glorious valley, what about Mr. Pinchot and Mr. Garfield, who gave it away without going from San Francisco to see this 'immediate jewel' of nature?"

Johnson concluded:

> Against San Francisco are thousands of Californians. The press of LA and Pasadena is in full opposition. A dozen associations, with headquarters in Boston, New York, Harrisburg, Chicago, Portland, Seattle, and San

Francisco, including the National California Association of New York, have passed resolutions against the desecration. Let this good work go on, so that the hands of California's grand old man John Muir may be upheld in this fight for his imperial state, for the whole people, and for future generations.

In turn, the mayor of San Francisco and his contingent questioned the character of Muir and his supporters, calling them "sentimentalists," "poets," and "Nature fakirs."[83] The San Francisco engineer Marsden Manson wrote that the preservationists were largely made up of "short haired women and long haired men."[84] In a letter to fellow members of Congress, California representative William Kent wrote,

> I hope you will not take my friend, Muir, seriously, for he is a man entirely without social sense. With him, it is me and God and the rock where God put it, and that is the end of story. I know him well and as far as this proposition is concerned, he is mistaken.[85]

For Johnson, these ad hominem attacks were merely symptoms of the problem the Yosemite Park was designed to solve:

> Cant of this sort on the part of people who have not developed beyond the pseudo-"practical" stage is one of the retarding influences of American civilization and brings us back to the materialist declaration that "Good is only good to eat."[86]

A frequent charge by both groups was that the other side was engaging in the "wrong" kind of publicity, namely, the manufacturing of public opinion.[87] To this charge of fiction versus fact was the antagonism of idealism versus practical reason. And by extension, the antagonism of Muir versus Pinchot. Spiritual lobbying in the name of an ethical obligation was no match for the disciplined and consistent information management of Pinchot's bureaucracy. Pitted one against the other, the spiritual lobbying for the forests could not match the ability to see like a publicist. The opposition was taken up and magnified in the place that mattered most for the constitution of truths in this era. To a large extent, the battle over these competing visions was a contest to articulate what the public meant in a democracy via the technology of legitimacy known as the popular press. Who this public was and how it ought to be constituted was of the utmost importance. While for some, this was about inventing ways to be and act together, for others, publics were made through the ordered acceptance of persuasive claims. While both Pinchot and Muir used the press to promote their visions as endemic to the public interest, it was Pinchot's promotion of the reservoir as an equitable distribution of resources—"the greatest good for the greatest number"—that captured the public's imagination.

At the same time, we must also recognize the groundwork Pinchot laid during his time as head of the Forest Service. The triumph of the Forest Service as the harbinger of truth lay partly in its ability to create facts through its management of information and partly in its vestment of power in the figure of Pinchot himself, whose political network of influence extended throughout Washington and nationwide. Making the truth is also a matter of making it harder to hear alternative versions of possible realities.

In a way, the battle over Hetch-Hetchy was not the beginning but the end: the culmination of Muir and Pinchot's competing efforts to wield influence in the public sphere. At the same time, Hetch-Hetchy brought about the invention of a new political concept, and with it, the embrace of the genre of advocacy required to wield that concept in American life. It was through Pinchot's exhaustive, strategic, and allied publicity that the new concept of environmentalism was decisively and authoritatively articulated.

2

Bringing the Outside In

Managing the "External Environment"

Making public relations into a legitimate profession had everything to do
with the objects around which the burgeoning practice coalesced: coal, oil,
steel, and rail. The technological and social transformations that enabled energy production on a mass scale gave rise to the political systems that grew
up around it. These political systems were not mere byproducts of the needs
of heavy industry; possibilities for political action emerged from—or were
subsumed by—the infrastructures created.[1]

In the first half of the twentieth century, public relations gained its footing
on the backs of these systems of mass production.[2] Yet the relationship between the infrastructure of industrial production and the infrastructure of
publicity is not as simple as it may initially appear. We tend to think of the relationship as one of mutual expansion and even co-creation. To some extent
this is true. As industrial production increased throughout the early 1900s,
so did the need for industrial public relations, which strove to make sense of
these industries for the range of communities whose lives and livelihoods
were affected by their output: workers, shareholders, journalists, political
decision-makers, and ordinary citizens. To support the material structures of
industry, public relations produced cultural structures of advocacy, helping
to make corporate operations visible as part of a broader social environment.

But as historians have shown, different industries had different political
objectives. And this affected the degree of visibility they sought and the cultural categories they used for people to understand them. In some cases,
making industry meaningful to its publics required a dissimulation of the
process of production. As Timothy Mitchell has argued, for instance, coal
mine workers were able to wield political power by mobilizing at localized
points of production in coal mines and sabotaging coal transport through
strikes, revealing the vulnerability of the energy-producing infrastructure
and drawing attention to workers' calls for voting rights, better working
conditions, and new political parties. Indeed, "between 1881 and 1905, coal

A Strategic Nature. Melissa Aronczyk and Maria I. Espinoza, Oxford University Press. © Oxford University Press 2022.
DOI: 10.1093/oso/9780190055349.003.0003

miners in the United States went on strike at a rate of about three times the average for workers in all major industries."[3] Strikes and other forms of sabotage of energy supply were crucial ways of making visible not just the labor itself but also the industrialized society's deep reliance on coal (and related industries of steel and railway lines) to work, travel, and live. Threatened with an interruption to their increasingly settled ways of life, public citizens' outcry swiftly prompted political response to workers' strike demands. Public relations for this industry therefore aimed to minimize the effects of worker strife by creating material and symbolic means for companies to demonstrate they were protecting their labor force and contributing to the public good.

This collective power of laborers was dispersed in the transition of energy supply from coal to oil. Structured via subterranean pipelines over greater distances and attended to by often migratory and temporary workers, there were few opportunities for workers to organize or counter the forces of industrial capital.[4] As chapter 3 will show, public relations helped make oil visible to its publics in a very different way: by reorganizing the concept of energy as a scarce resource in need of its own protection and making labor unions into allies in this endeavor.

The work of public relations in this era was therefore about something much more complex than merely promoting positive attitudes toward industrial operations. While in some cases PR was limited to creating publics that appreciated the material benefits of industrial production, in others, PR produced information that distracted or redirected publics away from the realities of production, or rendered production invisible altogether.[5] In all cases the goal was to create what industrial PR counselors and company managers called an "external environment" of acceptance for continued operations.[6] While the natural environment as a social and moral problem would not be named until the 1960s, PR in the interwar era took the broad terrain on which companies operated—social, material, and ecological—as valuable resources supporting industrial operations.

Most of the examples drawn on by business historians show how communications "counsel," in the industry parlance, was called in during moments of crisis faced by a powerful family, corporation, or industry.[7] For it was in these moments—heavily mediated, politically tense—that PR itself acquired visibility as a professional practice. This history is also at the heart of the critical assessment of PR agents as masters of spin. Although it is useful to look at these moments of crisis as nodal points in the industrial networks of legitimacy in American life, it is rather more consequential to recognize the persistent efforts by strategic communicators to manage the conditions of acceptance of industrial operations over time. For it is through the regular and ongoing forging of relations between extractive industries and their publics that it became impossible to imagine American democratic life without them.

In this chapter, we consider the ways that industrial public relations created infrastructures of advocacy to accompany infrastructures of mass production during the formation and embedding of what corporate owners called "industrial democracy" in the United States.[8] While chapter 1 examined publicity as a genre mediating emergent understandings of nature's role in the Progressive era of democratic reform, here we see how publicity becomes embedded in common practice as a professional system of collective representation. Seeing how the technology of public relations creates long-term *structures* of advocacy instead of merely devising messages in a crisis helps us to understand how forms of knowledge and information as well as strategies of representation were crafted and circulated among different industries. We also can see in this process the foundations being laid for industrial PR agencies as regular external counsel to companies, in addition to the consolidation of the PR function within firms, trade associations, and other industry coalitions or networks.

Considering PR as an infrastructure rather than as spin contributes to a broader awareness of the interconnection among information, the environment, and its publics in a modern democracy. The argument put forward here is that corporate PR did much more than make the environment over in industry's image; it made the material and symbolic infrastructure of mass industry visible as part of a system of democratic representation, by producing information that operated in particular genres and formats and creating publics oriented around certain political concerns and not others. Indeed, some of this making visible involves making invisible, through distracting, redirecting, or actively suppressing knowledge. Managing the "external environment" in this time period consisted of precisely this dissimulation. Industrialists wanted to "bring the outside in" through a variety of means, emulating democratic structures of advocacy by making workers into collaborating publics and beneficiaries of corporate governance.[9] In this way, dubious decisions by company leaders around environmental issues could be reframed as the outcome of collective agreement over what was best for all parties. Unsurprisingly, these decisions would have

consequential effects on the vital relationship between people and their natural environment.

We tell the stories here of two PR practitioners and their work to shape
the "external environment" for their industrial clients: Ivy Ledbetter Lee
and John W. Hill. In the pantheon of names and deeds associated with the
history of public relations, these stand out as strong representatives of the
more notorious practices associated with the profession. As we will see, this
is partly a matter of the way the histories were written; and in this sense we
tell their stories not to reproduce the ruts of past tellings but to provide an
alternative perspective on their influence on the character of environmental
knowledge. But it is also a matter of the outsized role each played in shaping
social and political contexts to make them favorable to industrial expertise.
While Lee and Hill are far from the only ones involved in the communicative
work to promote an extractive energy system, a focus on these individuals
and their networks of clients and allies gives us an inkling into the hierarchies
of promotional power that altered the possibilities for democratic claims in
industry's "external environment."

The Progress of Publicity: Ivy Lee and the
Machinery of Industrial Democracy

The historical legacy of the public relations man Ivy Ledbetter Lee comes from
two sources, each somewhat problematic. The first source is the twentieth-
century hagiography of public relations in general and of Lee in particular,
which paints the strategist as a scion of democracy, descended from a long
line of American figures preoccupied with ensuring the greatest good for the
greatest number of people.[10] His primary biographer, Ray Hiebert, positions
Lee in a genealogy stretching from Abraham Lincoln to the early American
settlers, on through Samuel Adams, Thomas Jefferson, and the illustrious
authors of the *Federalist Papers*, each reflecting the great American preoccupation with the rule of democracy as the direct will of an autonomous group
of individuals freely acting as a public.[11]

The second source, antithetical to the first, is the writing of political theorist Jürgen Habermas, who cites Lee as a key actor in the "refeudalization"
of the public sphere. Habermas calls Ivy Lee the father of PR, the mediator
of private interests in the public sphere, and the master of "staged public
opinion."[12] As Habermas would have it, Lee taught his industrial clients how

to "engineer" consent among different parties, which is inimical to the "time consuming process of mutual enlightenment" required for "a rational agreement between publicly competing opinions." For Habermas, this correlation was a dire manifestation of the closure of the public sphere to true representation of public interests. Instead, "privileged private interests" have "transmuted" the traditional notion of publicity—creating an object of public interest around which "a public of critically reflecting private people freely forms its opinion"—into the self-management of reputation in the pursuit of political power.[13]

Adherence to one or the other position is largely a matter of political leaning and is not usefully sorted out in a way that would satisfy both camps. Regardless, what is common to both portrayals is the emphasis on Ivy Lee's substantial role in shaping and influencing the progress of publicity in American democratic life. During his career Lee represented "nearly every facet of big business both in America and abroad: public utilities, banks, shipping, coal, oil, metals, sugar, tobacco, meat-packing, breakfast cereals, soap, cement, rubber, chemicals, investment companies, broadcasting, motion pictures, foundations, universities, charities, religious activities, political candidates, and the capitalists themselves."[14] He founded and advised industry and trade associations, in a spirit of what he thought of as intra-industry cooperation as well as a means to align his clients' efforts with the emerging standards for industry set by government regulators.[15]

In his time, Lee was renowned among industrial leaders and scholarly communities as an expert in matters of industrial representation. Born in 1877 to a Methodist family near Cedartown, Georgia, Lee graduated from Princeton University with a degree in economics in 1898. After a stint as a newspaperman in New York City, he moved into press agentry, becoming a publicist for local electoral campaigns and for financial investors; this helped him identify opportunities to do more expansive publicity work for corporate clients. As his accounts grew larger and his knowledge of markets more developed, he traveled extensively to give speeches and lectures across the United States and Europe in business, academic, and public forums. According to his primary biographer, he authored no fewer than eight books, eighteen pamphlets, sixty-nine articles in trade journals, magazines, and newspapers, and seven unpublished manuscripts. Each of these manuscripts elaborated the concept of publicity and its correlation to democracy in an industrialized society.[16]

One of Lee's most important legacies lay in the kind of work he coordinated across energy-related industries. The large-scale use of fossil fuels in the late nineteenth and early twentieth centuries created a situation whereby "a large majority of people in industrialized countries became consumers of energy generated by others."[17] At the outset of Lee's career, nearly all of his clients were public utilities—railroads, electricity companies, public transit—and the infrastructural providers of energy to support them: coal mine operators, shippers, and steel makers.

Lee's efforts to make visible the public purpose of heavy industry relied on promoting a national consciousness of consumer and political reliance on energy and its infrastructural requirements. As we saw in chapter 1, the emergence of a national public in the early twentieth century fomented national concern over abuses of power by monopoly interests, leading to multiple calls for reform. These calls for change extended in no small way to the public utilities, with public campaigns for government ownership of the companies and the land they occupied. The intensive consolidation of public utilities (over 3,700 individual companies were eliminated through mergers between 1919 and 1927) made individual companies seem to "disappear" before the public's eyes.[18]

By imposing a machinery of publicity onto the machines of industry, Lee returned visibility to the infrastructure of energy. Lee's idea was to bring energy to the fore in a way that signaled public and political participation in industrial decisions and to create a social and political environment in which industrial power was in the direct service of the public interest.

Two events in particular demonstrate Lee's agility in reimagining energy production for various publics. The first was the push by the railroads for a freight rate increase in 1913–1914. Since 1906, Lee had been publicity expert for the Pennsylvania Railroad. He worked on other railroad accounts as well—the Delaware and Hudson, and the Harriman lines: Union Pacific, Southern Pacific, Oregon Railroad, and Oregon Shortline—and helped found and advised the Association of Railroad Executives and the Bureau of Railroad Economics.[19] Much of Lee's initial work for the railroads had involved propping up their legitimacy in the face of declining public favor. Not only was railroad production decreasing as alternative forms of transportation appeared, but a culture of corruption among other industrial producers in the form of favoritism, kickbacks, and bribes had caught the attention of reform-minded regulators. The Interstate Commerce Commission (ICC) had slowly begun to put an end to those practices, establishing tighter regulation of the public utilities. Railroad executives had mainly responded by suppressing as much information as possible from public audiences, hoping that the low profile would protect them from further opprobrium.

Lee had a different idea. Just as the coal miners at the turn of the century had made visible the energy apparatus by controlling the flows of energy, so did Lee seek to make visible the energy apparatus in a different light. Lee exercised a parallel power, turning the railways into not just an energy machine but also an information machine that could produce alternative publics to evince greater support for the railways' cause. Relying on the notion that "publics do not exist apart from the discourses that constitute them," Lee used different kinds of information to produce platforms of debate upon which people from different walks of life could assemble and express support for his clients.[20]

To help his clients promote the freight rate increase, Lee made use of an eastern railroad publicity bureau in Philadelphia and another at the Railway Business Association in New York, which created ads in trade journals, issued circulars and pamphlets, and hosted journalists to whom railroad executives gave exclusive interviews. Lee and his team also wrote news editorials and articles giving reasons for the proposed rate hike in both large metropolitan dailies and smaller, local publications, reaching somewhere in the vicinity of 22,000 news outlets.[21] Not satisfied with the relatively limited representation of information in news organs, Lee wrote and mailed leaflets and bulletins directly to what he called "leaders of opinion"—"congressmen, state legislators, mayors, city councilmen, college presidents, economists, bankers, writers, lecturers, and clergymen," among others. To reach everyday passengers, he also posted bulletins in railway stations and left information folders in passenger railway cars, yoking the experience of train travel by ordinary individuals to the larger issues being lobbied in Congress.

Lee's campaign organized meetings with select audiences—chambers of commerce, boards of trade, and business clubs—and encouraged attendees to write letters expressing support for the rate hike to their associates in chain-mail fashion (ten letters to ten people) as well as directly to the ICC and to other White House officials. To cover more national territory, Lee coordinated a speaker's bureau, sending his clients and their representatives across the US to give talks in front of community groups, chambers of commerce, and boards of trade.[22] The audiences for these speeches were also asked to write letters of support, and the text of both the speeches and the letters was reprinted in the media. Just as Gifford Pinchot had done for state forestry, so did Ivy Lee create and coordinate a remarkably unified campaign for the railways, with each piece of publicity corroborating and reinforcing the claims of the others.

One way that communication on behalf of the railways brought the external environment into its ambit was to promote the function of the rail system in giving access to natural resources for the economic well-being of the local community. The *Daily Globe* of Joplin, Missouri, printed on 27 February 1914 a speech delivered at a banquet for local business leaders:

> There are now seven railroad systems entering Joplin; terminals are ramifying in every direction into a territory rich with all the products of the earth necessary to supply the wants of mankind. . . . Within a radius of 100 miles about Joplin are great forests of hard wood; untold wealth in coal, oil, gas, and stone, and in her fields the cotton of the south and the wheat of the north are neighbors, while the surrounding hills are famous for their small fruits and are known as the home of the big red apple.[23]

Other promotional media engaged in environmental boosterism, portraying the railways as the provider of "nature's metropolises."[24] Emphasizing the "special nature" of the middle west railroads, one document explained:

> This territory was created largely by physical conditions and natural development and growth of population. Its numerous lakes and rivers, some of which form its boundaries, attracted to their borders the early settlements and cities, and in time came Chicago and St. Louis, its two chief centers of population. The territory is comparatively level, which facilitated railroad building, and as population increased railroads were constructed in all directions, the objective points being naturally lake and river cities.[25]

The circulars and pamphlets also made use of a wide array of technical data, prepared and presented by the Bureau of Railroad Economics. The bureau, situated in Washington, was dedicated to the study of national and international rates, compiling statistical information and publishing bulletins for public use.[26] It was not well known at the time that Ivy Lee had founded and regularly advised the bureau. His command of railroad economics and his role as advisor allowed him to direct the presentation of information to promote his cause. Railroad building and transport statistics, census data, and economic ratios showing gaps between operating expenses and earnings were combined to demonstrate the need for rate increases by showing

US population growth, expanding miles of track laid, and increased railway traffic over a short time period.

Other bulletins emphasized the international situation, pointing to freight rate increases in France, England, Italy, and Belgium as justification for a similar move in the United States, or the risks to American reputation as a business powerhouse if other countries' industrial infrastructure became stronger. Still other documents considered the declining cost of domestic food and furniture, suggesting a rise in the living conditions of workers and an increase in their purchasing power, while no concomitant benefit was awarded to the rail companies engaging their labor.

Insisting that these data represented "not tendencies, but facts—now,"[27] Lee and his clients urged their audiences to take the measure of the benefits of the railroads as a public service and to submit their own facts in their letters of support, showing how the rate increase would affect their lives and livelihoods:

> We, makers of books to teach and educate the people, want to say that unless the railroads are not only treated fairly but liberally we can not prosper. Surely the whole people want it and beg for it, from ocean to ocean.—H. E. Smith, The Authors' Club, New York, 31 March 1914.

> We handle four to six hundred cars of grain per year. We have considerable trouble getting good equipment and urge that the railroads be granted 5 per cent increase effective on publication that they may purchase equipment and serve the best interests of the general public.—Robinson & Co., Lima, Ohio, 9 April 1914.

> The National Hay Association feel that while they are bearing more than their share of the burden with hay in fifth class, that for the best interests of the country in general it would be advisable to allow this increase of freight rates on all commodities.—H. H. Driggs, Chairman of Transportation Committee, Toledo, Ohio, 14 October 1913.

> Am a grocer of this city; have been here 26 years and grown gray in the service. . . . [T]he railroads are entitled to a 5 per cent increase in freight rates. That is brief and to the point; my reasons are those of 90 per cent of humanity.—M. M. Gasser, Duluth, MN, 14 January 1914.

We have these letters for present-day review because they were printed, as part of approximately 300 pages of additional evidence, in the *Congressional*

Record at the request of Senator Robert M. La Follette. La Follette, dubbed a "commercial Savonarola" by his enemies for his unwavering commitment to civic values over the base selfishness of industrial actions, was a member of the Interstate Commerce Commission and a staunch critic of Lee's scheme. Appalled at the scale and scope of the campaign to publicize the rail rate increase, La Follette insisted that he

> should have read into the Congressional Record every line and paragraph and page of this great mass of material to demonstrate the conspiracy that has been on in this country. It shall go, sir, to the people of this country a monument of shame, not only to those who would seek by that infamous method to control judicial functions, but to the press that lent itself to the imposition upon the public of this ex parte and unsworn mass of special pleading on behalf of the railroads.[28]

Dedicated to revealing the pathways of influence by which public opinion was "manufactured," the Wisconsin senator painstakingly represented Ivy Lee's publicity campaign in diagram form, adding this into the record as well (figure 2.1).

Deliberately absent from the diagram is the category of "the public." For La Follette, the campaign was nothing more than political lobbying by single-minded business interests to influence legislation. And at one level, that is exactly what it was. Assembling voices from a vast array of social, political, and economic organizations, Lee had harmonized the objective, the message, and the object of influence.

What the diagram does not show is that Lee's massive publicity campaign made visible the railroads as part of a much larger system, tied not only to political decision-making but to related industries and business concerns as well as to broader contexts of geography, natural resources, social conditions, and international trends. By creating the rail lines as products *of* and *for* their external environment, Lee had, by some accounts, succeeded in demonstrating the public utility of such a service.

On 16 December 1914, the ICC allowed the increase in railroad rates. The experience would serve to shape the notion of the public in this period. Writing in the *Electric Railway Journal* some years later, Lee would attempt to characterize the campaign as a true expression of popular will, asserting, in trademark style, that "the ultimate fountain of power in a democracy is, and must be, the people":

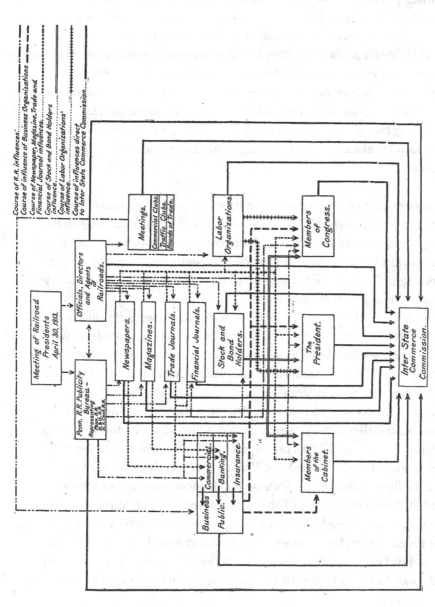

Figure 2.1. Diagram Showing Influences Bearing on Interstate Commerce Commission, 1914. Prepared by US Senator Robert M. LaFollette, this diagram was intended to show, as he put it, "exactly the extent and scope and perfection of this monster organization to control a decision" in the case of the railroads. *Source:* US Congressional Record 51 (1914), 7733–7734.

Newspapers, advertising men, and all interested in the progress of democratic institutions—whose ultimate safety must depend upon a fully
informed public opinion—should omit no opportunity to make it clear
to public officers, commissions, even Congress, that the people want to
know. It should, of course, be made equally clear that no one by aggres-
sive publicity methods or by extensive advertising campaigns can expect
to secure support for an unsound position. But it should be made so plain
that no one can misunderstand that any interest—public or private—
which earnestly, sincerely and candidly takes its case to the people shall
have strong public support for that fact if for nothing else. In other words,
every man is entitled to a full hearing, to his day in the court of public
opinion.[29]

The second event that allowed Lee to reshape the concept of the external environment would cement his notoriety—and that of public relations—in the
American imagination. In terms of information production, it was similar in
scale and in kind to the informating of the rail system. But its approach to the
objects of its influence—the workers at the mines—led Lee to test his theory
of popular will as a font of democracy.

On 20 April 1914, striking coal miners in Colorado were caught up in a
gunfight with the state militia outside the Ludlow work camp. Four were
killed immediately and eleven more were found dead in a pit beneath the tent
colony, suffocated by fire. An upsurge in violence across the coal fields in the
following days caused more deaths.

The news media was aflame with the story. The major target was the
magnate Rockefeller family, which owned major stock in the Colorado
Fuel & Iron Company (CF&I), the largest of 170 coal operators in the state.
Walter H. Fink, the Colorado Mine Union's publicity director, took up the
phrase "Ludlow Massacre" to describe the incident, and this was the label
applied to the event in most of the news coverage.[30] Public protests and
demonstrations followed, and a union-sponsored delegation visited the
White House, engendering more news coverage still. It was in this context that Ivy Lee was called on to stanch the public outcry against coal
operators and especially against the reigning symbol of industrial power,
the Rockefellers.

Lee immediately set to conducting extensive research to gather as much
information as possible about the mines. He dispatched his brother-in-law,
Lewis S. Bigelow, to Colorado to observe some of the mines and to collect

original documents that might serve Lee in formulating a response. As the public relations historian Kirk Hallahan recounts, "Materials collected by Bigelow included data compiled for government investigators, testimony, statistics on accidents, circulars sent to superintendents, maps, earnings data, correspondence, speeches by strike leaders, service records of superintendents and foremen, wage summaries for each mine, a brief history of the strike supplied by mine superintendents, and copies of the company's Camp and Plant house organ published from 1902 to 1904." Meanwhile, Lee himself collected mine statistics, organized a system of media monitoring to offset bad press, and conferred extensively with the Rockefellers about campaign possibilities.[31]

The best-known effort to come out of this data collection and monitoring was a series of nineteen bulletins, circulated by the thousands (11,000 in the initial print run, but the mailing list was eventually expanded to 19,000 names) to so-called leaders of public opinion across the country. The series title, "The Struggle in Colorado for Industrial Freedom," pleased the Rockefellers enormously. Shipped in bulk to the CF&I in Denver, they were then printed with the coal operator's headquarters address, making it seem as though they came directly from there. Eventually the first fifteen bulletins were reprinted into a booklet known as "Facts Concerning the Struggle in Colorado for Industrial Freedom" and recirculated to 40,000 additional people.[32]

But the true transformation of information management came through outreach to the workers themselves. One of the ways that worker strikes had become effective tools of leverage was because cutting off energy supply at chokepoints made immediately and dramatically apparent to ordinary people the importance of the mine workers' output to the everyday tasks on which their households and work depended. Lee attempted to reverse the effect of the strikes on public opinion by mediating the relationship between employers and their labor force. Lee's efforts were supported by another advisor hired by the Rockefellers in 1914, Canadian labor expert and future Canadian prime minister W. L. Mackenzie King, to undertake a study of labor and capital relations and determine the so-called corporate responsibility of the coal operators.[33] Mackenzie King's recommendations were to develop an industrial plan, one that would offer workers forms of representation within the company. Rather than letting the miners represent the industry to the public, Lee resolved to put the coal operators in front of the public, showing how well the miners' demands were attended to. He then created public events in which CF&I managers would engage with workers at their places of work.

This was hardly a challenge in the Colorado coal mines. CF&I miners lived in company houses in company towns, attended churches and schools owned by the company, and bought supplies at company stores.[34] Welfare programs had been in place for the CF&I since the nineteenth century, with medical and social services such as a hospital, employees' clubs, cooking classes, and music groups.[35] The innovation of the Colorado Industrial Plan was to further embed the lives of workers into the infrastructure of industry.

Leveraging his ties to the railroads, Lee collected further intelligence about union officials on labor conditions in Colorado.[36] He then conducted his own visits to several mines: Primero, Segundo, Frederick, Sopris, Morley, Tabasco, and Berwind, and visited tent colonies at Starkville and Ludlow.[37] Armed with firsthand site details, Lee next crafted a charm offensive for Rockefeller Jr. He set the stage by arranging for leaflets containing the CF&I's employee representation plans to be sent directly to workers' homes so that they and their families would see these plans in a domestic setting. He also had posters placed *at* the mines to reach workers in their places of work. One poster said:

> It is the purpose of our Company not only to pay high wages, but to make all other conditions of employment satisfying to our men.
> We want every man who works for the Colorado Fuel & Iron Company to feel that the Company is his friend.
> We will at all times be glad to have you send us, in writing, any suggestions which you may feel will advance your own welfare, that of your fellow workers, or that of the Company.
> We want every man to be happy in his work, and we hope you will help us to make you so.[38]

Finally, he sent Rockefeller Jr, along with Mackenzie King and a team of reporters, to glad-hand with the workers in their homes, ask questions of their wives, and visit CF&I facilities, including housing and a school.[39] Such PR events were closely covered by local journalists in addition to the national coverage they received.[40] By dispatching company information, representatives, and journalists to key sites at the mines, Lee's campaign helped to reimpose industrial power over strike power, defusing the coal mining infrastructure as a site of protest.

Organizing on the Inside

The key to understanding the employee representation plan (ERP) as a form of publicity is to observe the way it served to organize the inside and the outside of the energy-producing companies that adopted it. After CF&I workers voted to approve this form of representation, Rockefeller was lauded in the media for his reform-minded leadership. The "Rockefeller Plan" quickly spread to other companies threatened by labor unrest, in oil, steel, electric utilities, and heavy machinery. Sociologist G. William Domhoff writes that this cross-industrial spread was enabled by Rockefeller's broad stock ownership: it was Rockefeller who pressed the other companies in which he held stock to adopt employee representation plans.[41]

On the inside, these "company unions," as they were known, were promoted as a site for the practice of democracy. Within the plants, mills, lines, and mines, the ERP provided workers with deliberative councils, representative elections, and participation in major decisions taken by the firm. This company union bore the appearance of rights to collective bargaining for workers in support of claims to autonomy. Yet the existence of a seeming democratic structure within the walls of the company's property in no way diminished the authority of its owners. "The employer retained unilateral control over final decisions, manipulated elections to ensure compliant employee representatives, and blocked intra-union communication by which the rank and file could form its opinions and monitor and instruct its representatives."[42]

The ERP served as a legitimating device—an object of compromise. It created a new form of coordination and collaboration between workers and their owners in an arrangement known as industrial democracy. Worker strikes had made visible the socio-technical infrastructure of energy production, which gave workers a political power they had not previously had. Industrialists dissimulated this infrastructural power by bringing the outside in. By turning labor relations into an internal problem, one that could be resolved within industrial walls, companies could recognize their workers on company terms, safe from external unrest as well as political and legal scrutiny. As the vice-president of the Republic Steel Corporation would put it in 1939,

> The gospel of better understanding between men and management . . . has
> reached practically every worker in Republic through their representatives

and through proper administration of industrial relations by the foremen. The most important labor relations job is that of making men better acquainted with management. *It can only be done on the* inside *of your plant, and there it must be done so thoroughly that it cannot be undone by those on the outside.*[43]

Indeed, this renewed internal authority was designed to manage both immediate and longer-term risks posed by the "outside." In the immediate context the outside consisted of labor organizers, government policies, and emerging laws and codes that awarded rights to workers in the context of industrial collusion and unfair competition. More broadly, the outside was also the natural environment, which was at this time little more than an obstacle to industrial growth. Industrialists fought desperately in this period to maintain the power of industry as the engine of progress in the American mind. If the industry of energy was to continue to function as "the touchstone of our fortunes and the barometer of our condition," company leaders would have to work not just to create internal homeostasis but also to manage another "external environment"—the terrain of public opinion.

A New and Sturdy Civilization:
Hill & Knowlton as Spokespeople for Steel

In the face of continued attempts by government and labor organizers to transform industrial operations, companies moved to manage their external environment through a deliberate and expansive program of public information. Employee representation programs were joined by statistics, studies, surveys, speeches, and civic events in massive quantities and used to promote the industry to its relevant publics in the form of news releases, booklets, pamphlets, and radio broadcasts. It was hoped that promoting the internal activities of firms would serve as a bulwark to the external efforts to control industrial production.

The US steel industry is representative—but not exclusive—in this regard.[44] By the end of the nineteenth century, the United States dominated the global output of iron and steel. Integrated technological production, low rail tariffs, and superior lake transport and shipment organization dramatically impacted the economic and infrastructural development of the country.[45] Backed by industrial titans J. P. Morgan, Andrew Carnegie, and Charles

M. Schwab, steel was part of the modern industrial order. This order combined mastery of nature through the manipulation of industrial materials with another kind of mastery: influencing the public and political accommodation of the climate of industrial capital. For some of the most prominent representatives of the steel industry, this would come about through a focus on publicity. As US Steel Chairman, Judge Elbert H. Gary, told *Harper's* magazine in 1908:

> When the Steel Corporation was formed we proposed to give frank statements to the public regarding our affairs, telling what we were doing and what we purposed doing. There's no sense in being blind to the times. Everybody has got to go to work and get straight and stay straight, and the thing most essential to that is publicity. I believe in it, first, last, and all the time. By publicity I don't mean advertising. We don't need that. I mean letting the public know what you are doing and how you are doing it, so long as the management of your business is legitimately a matter of public interest, in order that reassurance may grow and confidence may be maintained.[46]

The American Iron and Steel Institute (AISI), founded by Gary in 1908 as a likely response to the 1907 financial panic, gives us an object lesson in the industry's particular interpretation of publicity in the course of its expanding influence over the external environment. Like its forebears (the American Iron Association, founded in 1855; and the American Iron and Steel Association, founded in 1864), AISI was an information hub "for the mutual interchange of information and experience, both scientific and practical." AISI's role was to collect statistics on trade, maintain a library of trade-related publications, and promote education for apprentices.[47] It was in the 1930s that AISI would begin to develop a proper information infrastructure to overlay its industrial one.

Starting in 1933, AISI was charged with a more prominent role in industrial publicity. In June of that year, New Deal efforts led the US Congress to enact the National Industrial Recovery Act (NIRA). This act, "a unique experiment in U.S. economic history," was an effort by government to enlist industry in supporting the rights of citizens as workers and consumers during the Great Depression.[48] Companies were enjoined to create industrial alliances in each sector, developing codes of conduct within these alliances to regulate prices, wages, and quotas for production. More consequential still for industry, NIRA gave employees the right to organize, collectively bargain, and join an external labor organization. AISI was tasked with developing the codes and standards for steel.

In hindsight it is not at all surprising that John W. Hill was hired by AISI almost immediately after the passage of NIRA. A onetime journalist, Hill had opened a "corporate publicity" office in Cleveland, Ohio, in 1927. In 1933, partnering up with Don Knowlton and opening Hill & Knowlton, Inc., he would establish what is today one of the largest public relations firms in the world. According to the historian Scott Cutlip, Hill's interest in public relations was stimulated by Ivy Lee. He came across a copy of Lee's 1926 book, *Publicity: Some of the Things It Is and Is Not.*[49] Lee had mused in this book about the challenges involved in giving the public facts about an issue. "To present a complete and candid survey of all the facts concerning any subject is a human impossibility," he began. He quoted Walter Lippmann, whose own views on the topic were contained in his 1922 book, *Public Opinion.* "The whole of public affairs cannot be reported, and in that simple, and rather obvious, but unappreciated fact lies one of the fundamental problems of public opinion."[50]

Making a distinction between the "absolute"-ness of truth and the connotative indeterminacy of facts, Lee concluded that the only way to put information in front of the public was to provide "my interpretation of the facts."[51] This is not propaganda, he stressed; or at least, not the negative associations of the term. The potential taint of propaganda can be avoided by the public's use of judgment in making sense of the information received. Was it not a fundamental democratic principle, he argued, to give each person the right to judge and evaluate the value of facts for themselves?

Hill took these ideas to heart. His first task for AISI was "to get it established as a recognized source of trustworthy information about the steel industry."[52] He began to publish a regular bulletin titled, of all things, *Steel Facts.* "These publications," he wrote, "provide one straightforward way of taking an industry's story directly to the public." Within eighteen months, approximately 1 million copies of thirteen issues of *Steel Facts* had circulated. Hill's bulletins responded to the NIRA's requirement to establish industry standards and codes, publishing data on freight tariffs, labor statistics regarding wages and hours, and specifications of commercial iron and steel products for workers, consumers, and company heads. It also aimed to increase the types and functions of information that made up the business of steel production. Increasingly, for instance, articles contained information about upcoming legislation that stood to negatively impact the industry and offered advice to managers on how to reckon with it.[53]

A second, critical task undertaken by Hill was to reorganize the topography of information management for steel. In 1935 he opened an office in Washington so that AISI could take a visible stand to oppose legislation it deemed problematic.[54] He created a committee dedicated to public relations within AISI and also established community offices in Chicago, Pittsburgh, and other locations in the East, proximate to important steel companies, to help them create information programs for employees and local residents. Over the next couple of years, he took on additional steel clients, Republic Steel and Midland Steel, as well as clients who supplied raw materials or machine tools to the steel industry. By establishing these information nodes on a growing network of public relations management, Hill helped steel preserve its "industrial democracy" from the incursions of outside antagonists.

Nevertheless, the strength of organized labor was growing. The National Labor Relations Act (also called the Wagner Act) was signed into law in July 1935, aiming to overcome the problems of enforcement under NIRA.[55] The leader of the Mine Workers Union, John L. Lewis, had publicly stated his intention to organize steelworkers, forming the Steel Workers Organizing Committee (SWOC) and creating, Hill recounted, a state of "near panic" among his clients.[56] Steel leaders turned to Hill to help them redouble their efforts.

New efforts included news releases by Hill & Knowlton on behalf of AISI, reaching at times 2,000 daily and weekly papers. The press releases used employee representation plans, now widespread among steel and other companies, as both sources of data and as media of publicity. Some announced recent "studies" of employee representation plans whose results showed substantial benefits of collective bargaining to the firm. Others announced results of internal elections held by steelworkers that showed a majority were opposed to strikes and favored ERPs over external organization. Still others declared that surveys of employees revealed active support for ERPs. This industry "data" was taken as factual evidence that companies protected employees, not only by awarding them voting and bargaining rights but also by shielding them from the "intimidation, coercion and violence" of John L. Lewis's attempt to unionize workers and from the long arm of the Wagner Labor Relations Act.

In addition to the press releases, Hill & Knowlton prepared complete news articles, offered free of charge to news editors in "an effort to present dispassionately the steel industry's position on collective bargaining and to explain this industry's refusal to accede to the demands of the professional labor union leaders."[57] The news releases and the articles also cropped up in *Steel Facts* in slightly different language to speak more directly to steel employees. Using scripts written by Hill & Knowlton, industry representatives appeared on radio shows and gave interviews. New booklets were printed and distributed in large numbers. *The Men Who Make Steel* emphasized the "practice of cooperation" between labor and management in the steel industry and suggested that "since labor seeks a wage from industry, management [seeks] a salary, and capital seeks a return on its investment, the aims of each group are identical."[58]

More than merely aligning workers with the company, these publications worked to articulate steel as a nodal point on a network of modern civilization:

> Steel has defeated time and distance. Steel forms the frame of our mighty and beautiful buildings, the skyscrapers and the factories, the massive public offices, the railroads and the subways. Steel has made it possible to erect upon this continent a new and sturdy civilization, which has freed man from the back-breaking, soul-consuming toil that characterized the life of his ancestors.

This civilized self was, the publications took pains to remind their readers, supported by a skeleton made of raw materials and natural resources:

> Steel, with its billion of money; Steel, with its myriad glowing furnaces, its thundering mills, and its smokestacks thick as stalks in a cornfield; Steel, with its thousands upon thousands of miles of ore land and coal land and gas land; Steel, with its endless railways and its fleets of vessels; Steel, with its swarming population of workmen and its trade lines penetrating every business and every corner of the world, has become the touchstone of our fortunes and the barometer of our condition.[59]

In 1937, having finally signed a union contract with US Steel, Lewis called an organizing strike against the so-called little steel companies, primarily Bethlehem, Republic, Inland, Youngstown Sheet & Tube, and Jones & Laughlin. These were called "little steel" only in proportion to the US Steel behemoth; in actuality they were large companies. Despite the concerted plan of information provision by AISI and its network, the causes of organizing workers were favorably portrayed in certain publications, such as the *Pittsburgh Courier*, *Harper's*, and the *Nation*, as well as in workers' magazines. But as law professor Ahmed White points out, the advantages of worker solidarity—the ability to communicate directly with members of local communities and their families—had no influence with members of the middle and elite classes, whose opinions shifted from indifference to opposition as pro-business publicity took hold.[60]

In May 1937, John Hill met with AISI members to discuss plans for defeating the strike. The list of tactics was long. Hill and the AISI solicited the financial and network support of the National Association of Manufacturers (NAM). The NAM had, since its founding in 1895, functioned as "the voice of the manufacturing industry in the United States." Its motto, "Industrial democracy in action," was put to use for the steel industry via a front group, the National Industrial Information Committee. This committee, according to NAM's promotional literature, was put into play in 1934 "to utilize every practicable media of communication to broaden the public's understanding of the private enterprise philosophy and to stimulate public resistance to attacks on the system."[61] This included a nationwide campaign promoting what the NAM called "harmony" between employers and employees: posters and advertisements with slogans like "Are You an American Citizen?"; "To the Leaders of Public Opinion"; and "Prosperity Dwells Where Harmony Reigns."[62] During this time, the NAM effectively became a client of Hill & Knowlton, disbursing monthly payments.[63]

Hill & Knowlton sought to rally other clients to the cause, a goal easily achieved because many of the clients on the PR agency's roster were either steel companies themselves or employers in related industries, all of which were opposed to unionization (table 2.1). In a feat of industrial management, Hill's innovation was to turn nearly all of these client organizations into additional nodes on his anti-union information network. With the exception of New York–based AISI, all of the companies were located in Ohio, already creating a local ecology of like-minded constituents. With each organization now serving an additional role of information sharing, public outreach, and collective mobilization in opposition to labor organizers and government laws, Hill now oversaw a powerful and multi-layered structure for the management of public opinion around the benefits of industry.

Table 2.1 Clients of the Public Relations Firm of Hill & Knowlton, 1933–1937

Hill & Knowlton Client	Type of Organization
1. Otis Steel Company	Steel company based in Cleveland, Ohio
2. Petroleum Industries Committee	Committee formed under the auspices of the American Petroleum Institute
3. Youngstown Sheet and Tube	Steel manufacturer
4. Standard Oil Company (Ohio)	Oil company
5. Pickands Mather	Supplier of raw materials such as ore to the steel industry
6. National City Bank (Cleveland)	Commercial banking
7. American Iron and Steel Institute	Trade association
8. Eaton Manufacturing Company	Automotive tools and parts manufacturer
9. Warner-Swasey Co.	Machine-tool maker
10. Republic Steel Corporation	Large steelmaker based in Ohio
11. Berger Manufacturing Company	Division of Republic Steel Corp.
12. The Austin Company	Factory building designers
13. Midland Steel Products Company	Manufacturer of steel products
14. The Block Company*	(Steel manufacturer)
15. Trundle Engineering Company	Cleveland-based consultancy
16. Euclid Avenue Association	Cleveland-based city planning association
17. Electric Vacuum Cleaner Company	Affiliated with General Electric
18. Retail Merchandise (Merchants) Board, Inc.	Merchants' organization
19. Cleveland Bakers Club	Trade group
20. Greater Akron Association	Industry group
21. Great Lakes Expo	Industry group
22. Akron Chamber of Commerce	Chamber of commerce
23. Cleveland Chamber of Commerce	Chamber of commerce

TOTAL RECEIPTS: $412,004.49

Source: US Congress, Violations of Free Speech and Rights of Labor: Supplementary Exhibits (Exhibit 6305), US Congress Hearings 76 Session 1 (1939), 15523.

* It is unclear to what this listing refers. Possibilities include the W. G. Block Company, a fuel merchant based in Iowa; or Inland Steel, a Chicago-based steel company founded and operated by the Block family (publications of the era refer to it as the Blocks' Company). Given John Hill's frequent reference to Joseph L. Block in his published work as well as the clear orientation of Hill & Knowlton clients in this time period toward the steel industry, we suspect it is the latter.

A few examples of this collective effort during the Little Steel Strike demonstrate their use of a "civic" rationale to justify their actions.[64] One was the
formation of "citizen committees." During the 1937 strike, Hill & Knowlton
staff worked with local chambers of commerce to mobilize citizens to act
on behalf of business in their communities.[65] These groups undertook
anti-union publicity campaigns under the sobriquets Citizens' Committee
and Steel Workers' Committee of Johnstown. While staffers went to strike
locations in Youngstown, Canton, Warren, Massillion, and Monroe, preparing and circulating news releases, Bethlehem Steel and the so-called citizens' committees hired the services of additional public relations agents: John
Price Jones Corporation of New York, which raised funds for newspaper advertising through a letter-writing campaign to local donors; and Ketchum
& Co., which handled radio publicity. The citizens' committees also funded
the preparation of blank back-to-work petitions. By the end of the strike, the
committees had expanded across the country, drawing in chambers of commerce from around seventy-three communities.[66]

A second initiative involved the use of another kind of third-party representative: journalist George Ephraim Sokolsky, syndicated Hearst colum-
nist for the *New York Herald Tribune*. Sokolsky ghostwrote news articles
and pamphlets on behalf of Hill & Knowlton.[67] Sokolsky also spoke at the
Cleveland and Akron Chambers of Commerce (both Hill & Knowlton
clients) as well as at " 'civic progress meetings' arranged and paid for by local
employers but publicly sponsored by 'neutral' groups." Sokolsky showed
his audiences statistical and other data provided by the Greater Akron
Association, among others. He also appeared weekly on a radio program
sponsored by the NAM.[68]

When, in 1939, the Senate Committee on Education and Labor formed a
special subcommittee to investigate the obstruction of free speech and collective bargaining during the strike, it made sense to appoint La Follette as
chair. La Follette did not hesitate to bring Hill & Knowlton to task, putting
into the *Congressional Record* once again a list of the mass of material generated by public relations efforts to control the external environment in which
industry operated.

Forty-two years after Hill was hired, AISI was still a major client of Hill
& Knowlton.[69] The agency's campaigns on behalf of the steel industry in
the decades after the Second World War adopted new media and market
research techniques, using more elaborate surveys, psychological studies,
and telecommunications networks to take the pulse of public and political

Table 2.2. "Informating" the Little Steel Strike, 1937. Documents prepared by Hill & Knowlton on behalf of the Republic Steel Corporation.

Unit/No. printed copies	Description	Circulation
77,000 copies	Four-page letters on labor policy from Republic Steel Corp. to employees	Republic Steel Corp. employees
100,000 booklets	"The Real Issues," listing reasons steel companies would not sign labor contract with CIO	Republic Steel Corp. employees. Mailings to "100 colleges and universities and to a total of 8000 investment bankers and dealers throughout the U.S., plus 650 copies to newspapers and 360 copies to public libraries"
500 copies	"Memorandum Governing Collective Bargaining"	unreleased
2,200 reprints	Editorial from *Daily Metal Trade*	1,900 newspapers
2,100 booklets	"Who Is John L. Lewis?" denouncing the labor organizer for his "ruinous class collaboration" and "opportunistic" methods	Employees across all 21 plants of Republic Steel Corp.; 200 to National Association of Mfrs.
3000 booklets	"What I Would Do to Maintain Democracy" transcript of radio broadcast by Republic Steel Corp. chair and president T. M. Girdler	
2000 copies	Transcript of address by T. M. Girdler at Warren Chamber of Commerce	
39,000 booklets	"CIO versus American Democracy"	

Source: US Congress, Violations of Free Speech and Rights of Labor: Supplementary Exhibits, US Congress Hearings 76 Session 1 (1939).

fervor.[70] But the strategies of action remained constant. In the making of the machinery of consent for the continued operations of steel and its partner industries, Hill & Knowlton maintained its themes of fair representation, collaboration, and the provision of the public good. This was not so much about creating "a voice for the industry" or even for intra-industrial interests as it was about integrating the public relations function into industrial operating structures. The ongoing need to legitimize the production of coal, rail, steel, and oil as sources of a civic American self would require the constant cultural shaping of these polluting industries by public relations activities.

Assessing Hill & Knowlton's infamous efforts to promote the tobacco industry starting in the 1950s, critics have pointed out the difficulty in sorting out where the PR firm ended and the tobacco companies began.[71] Looking at the PR for rail, coal, and steel over the first half of the twentieth century shows the extent to which this was true much earlier. Equally important is the way these industries were dedifferentiated in the public relations practice. While, as we have seen, the potential for political power by workers in different energy industries can be wielded according to both the specialized skills required for different labor practices and the infrastructure of production, the work of public relations relies precisely on overcoming these distinctions. In the systematization and industrialization of public relations itself, the self-conscious role of the communications professional is to bring external problems within the purview of companies and to make these problems appear simpler and therefore easier to resolve by informational means. Dedifferentiation of industrial energy sectors gave PR practitioners more than just access to a larger stable of clients. It helped create a unity of purpose, strategy, and message across them. This harmonization rendered the message more legitimate, authoritative, and omnipotent. The "external environment" public relations sought to manage consisted of a range of publics: journalists, company employees, legislators, association members, businesspeople, and civic leaders. In building an infrastructure of advocacy to overlay that of industry, industrial public relations of the era integrated information, environment, and publics to achieve legitimacy to such a degree that it becomes impossible to disarticulate them.

The interconnectedness of this triad makes it difficult to characterize public relations as exclusively a cynical exercise in the pursuit of selfish ends, as La Follette would have it. But despite the obvious indications of scale-tipping on the part of Lee, Hill, and their networks, the coordination of perspectives in the pursuit of a defined public good is more than a strategic maneuver. As the sociologist Lyn Spillman writes, evaluating PR's appeals as sincerity or cynicism is a false dilemma. Part of the problem is assessing just what the public good is supposed to mean, a topic we will address in detail in chapter 4. But part of it is also that it is impractical at best, and distorting at worst, to disarticulate the nexus of information, environment, and public in the social, economic, and political context of its making. Instead of focusing on "personal sincerity or guileful opportunism," Spillman suggests, the task is to assess "the vocabularies of motive" by which business associations and their representatives understand what they are doing and how they convey this to their audiences.[72] This is the task we apprehend in the next chapter.

3

Environment, Energy, Economy

The Campaign for Balance

It is ironic that the greatest single moment of visibility in the twentieth century of the effects of environmental hazards in the United States was prompted by a book about invisible toxins. Excerpted in the *New Yorker* magazine, which immediately galvanized its shocked and panicked readers, Rachel Carson's *Silent Spring* imagined a world without birdsong, a landscape ravaged by the unseen devastation of chemical pesticides. A former federal employee with the Department of Fish and Wildlife, a naturalist, and a longtime science writer, Carson's flowing prose was augmented by her credible and well-documented argument. Connecting government agencies with chemical industry irresponsibility and the collusion of academic scientists, her book had a damning effect on all three pillars of society.

It is an oversimplification to attribute the transformation in attitudes toward the environment in the 1960s to a single event. Yet it is difficult to overstate the impact of Carson's *Silent Spring* on the American public when it appeared in 1961. As William Sewell has demonstrated, events may bring about historical change by transforming the very cultural categories that shape human action.[1] *Silent Spring* gave Americans a cultural schema with which to coordinate disparate and until then largely unexpressed views of the natural environment. The book not only raised the alarm about the toxic hazards of pesticides; it also fomented a groundswell of public reform aimed at reclaiming the rights of the citizen to a safe, clean, and healthy environment. This reform movement set its sights on the output of private industry and the government's inaction to control it.

In its ability to raise popular consciousness and mobilize ordinary citizens for collective political action, *Silent Spring* offered its readers a new claim to democracy. It created a profoundly original kind of public—a public that recognized something called "the environment" as a fragile natural resource and could exercise its right to call for its protection. *Silent Spring* helped not only to hold chemical industries directly responsible for severely damaging the

A Strategic Nature. Melissa Aronczyk and Maria I. Espinoza, Oxford University Press. © Oxford University Press 2022.
DOI: 10.1093/oso/9780190055349.003.0004

quality of the public goods of air, water, and land, but also to show the social effects of polluting industries. The vision of a technologically advanced, affluent, and industry-centered future that contentious industries had worked so hard to promote in the interwar years was disappearing. In this regard, environmentalism opened up the terrain for political participation, as individuals developed together a new ethical orientation for an alternative, egalitarian, and more sustainable future.[2]

By all accounts, the industry's immediate response to the public outcry raised by *Silent Spring* was a disaster. Attempts to discredit Carson through damning book reviews, newsletter mailings, television appearances by "expert" scientists opposed to her findings, and letters to news editors questioning her credibility had the opposite effect of what was intended. If *Silent Spring* "created" the environment as an object of public concern, industry's attempts to quell the problem served only to reinforce their position as antagonists.[3] The public had been awakened to a concept of the environment as a public problem, and the source of the problem was placed squarely in the laps of industry.

The strength and character of reform embodied in the environmental movement was all the more exceptional considering how effectively industry had been promoted as beneficent in the previous decades. An essential dimension of public relations work in the years from the end of the First World War through the 1950s was to promote manufacturing industries and their suppliers as symbols of technological and scientific progress. Major companies in the steel, automotive, chemical, and electric sectors launched extensive PR campaigns in the 1930s and after to demonstrate their technological mastery and adherence to the latest scientific knowledge. Above all, companies wished to place themselves in the central role of providing an advanced quality of life for Americans. At world's fairs and exhibitions, through colorful imagery in advertising campaigns, and with an expanding array of consumer products, corporate PR linked industrial output to economic growth, and economic growth to the benefits enjoyed by a rapidly growing consumer class.[4]

Industries' reversal of fortunes in the 1960s was due, in part, to its weakening control over this favorable image. The "public relations craze" of the 1930s, as the historian Roland Marchand has called it, had been part of the ongoing effort to promote the value of free enterprise and to suppress labor agitation, continuing the projects of strategic visibility initiated by PR men such as Ivy Lee and John Hill and their affiliated networks, as we saw in chapter 2. By publicizing the value of industrial production for individual Americans' well-being as consumers and citizens, public relations tried to deflect growing complaints by residents about the effects of industry's output on their communities' air, water, and land.

In some instances, reports of local residents complaining about factory odors or "murky and unpalatable" drinking water were suppressed by the town's officials, who prioritized the economic benefits of the factory's presence in their community and reinforced the belief that "despite the inconvenience, dirty air and water was the price one paid for industrial prosperity."[5] In other settings, pollution was positioned as a source of scientific and technological progress in itself, with industrial advertisements depicting puffing smokestacks or the machinery of burning coal as symbols of American ingenuity and power.

Industrial producers had also made strategic use of scientific knowledge about the effects of their products. And this was easy, as prior to the 1960s, the source of scientific knowledge about the effects of pollutants on people and their environment was the industrial laboratory. Corporate labs had since the 1920s documented health problems and even deaths among employees in the workplace. Drawing on the emerging science of "industrial hygiene," corporate scientists developed strategies to mitigate "occupational disease."[6] Until the Second World War, the science of industrial hygiene was substantially dependent on industry funding. Since the research was underwritten by companies, studies of worker illness were documented in confidential reports to the sponsoring company rather than appearing in scientific journals; and findings were used to overturn lawsuits or worker compensation claims.[7] This, too, would change in the 1960s and after, as increased federal funding for scientific research transformed both the pool of scientists considered experts and the nature of regulatory science.[8]

This chapter is about how industry lost its grip on the narrative of environmental expertise, and how it got it back, helped along by a dedicated and wide-ranging campaign of public relations. We focus initially on the activities of the Manufacturing Chemists Association (MCA), the primary trade association for the American chemical industry. Founded in 1872, the MCA (renamed the Chemical Manufacturers Association, CMA, in 1978; known today as the American Chemistry Council) currently represents over 170 companies at all stages of chemical manufacturing, a business worth somewhere in the vicinity of $565 billion. In addition to examining the role of this trade association in the promotion of public relations expertise in the interwar period, we also consider how it impacted the politics of environmental control. To do this we focus on the activities of one of the most notable PR practitioners during this time: E. Bruce Harrison, whose decades-long career in "green" public relations—which began at the MCA—would set the rules of engagement by business leaders with environmental problems. Harrison's efforts to restore "balance," as he framed it, to industrial expertise to challenge the growing legitimacy of the US environmental movement left indelible marks on the terrain of regulatory possibility as well as on the modern conception of environmentalism.

To be sure, the MCA was not the only trade association involved in turning the tide of public perception back in industry's favor; nor did efforts emanate from the chemical sector alone. As we shall see, many other extractive and energy companies and their trade associations formed part of the industrial campaign for "balance" in the 1960s and after. By adopting a more targeted perspective on the chemical industry as well as on the incremental information and communication strategies adopted by E. Bruce Harrison, many of which stem from his work with the MCA in direct response to the publication of *Silent Spring*, we show how specific tactics and motives were initiated and reproduced across industrial terrain as well as in the political and public imagination.

Trade Associations and the Promotion of Expertise

Long before industrial actors were taken to task by the active intervention of environmentalists and regulators, companies recognized the potential impact of pollution on their business and developed clear-eyed—if distinctly compromised—strategies of organizational response. By the end of the 1940s, three trade associations—in chemical manufacturing, steel, and petroleum—had created research and information programs focused on air pollution.[9] There were two primary reasons for these trade associations' newfound interest in air pollution. The first was a public health crisis: a deadly smog that settled for five days over the town of Donora, Pennsylvania, in October 1948. A temperature inversion had trapped the smog near the ground, killing twenty-two people and sickening around 6,000 more. Media attention and a national inquiry in subsequent months identified a local company, a subsidiary of U.S. Steel, as the culprit.[10] Other companies in the steel industry as well as their peers in petroleum and chemical product manufacturing sought protection from their trade groups from any further fallout of the crisis.

But it was just such "fallout" that led to the second reason for trade groups' commitment to air pollution research: the Air Pollution Control Act of 1955, the first-ever US federal legislation involving air pollution. Calling it a "control" act was slightly misleading: the act only made provisions for research and funding for air pollution control rather than imposing any restrictions on industrial output (this would come in 1963 with the Clean Air Act). This gap between the mandate for research and the mandate for regulation was the space into which industry would insert itself.

Over the course of the 1950s, the Manufacturing Chemists' Association became one of the most authoritative sources of research and information about air pollution in the United States. It was an effort to control both the disease and the cure. Beginning in 1951 and continuing throughout the decade, the MCA produced a series of reports for officials at municipal, state, and federal levels. One reference standard was the MCA's *Air Pollution Abatement Manual*, which contained information on technical aspects of pollution abatement; terminology of common vapors, gases, dusts, mists, and fumes; the relation of meteorology to air pollution; legal issues and requirements; and principles of community relations, in addition to a section on physiological effects. A second booklet, *A Rational Approach to Air Pollution Legislation*, was a more normative document. It proposed an organizational structure for government and industry that assigned standard-setting and enforcement of air pollution rules to local commissions populated by industry members.[11] The MCA also maintained a pool of industry experts on air pollution control to work with officials involved in the preparation of legislation or regulations. These experts testified several times before Congress on legislative issues.[12]

The MCA's development of expertise and information was designed in part to ward off the growing concerns about industrial producers as polluters. It aimed not only to interact with government and scientific decision-makers but also, importantly, with the general public. In 1949, the MCA created a public relations committee, hoping, as one MCA member put it, "(1) to offset the adverse effects caused by the activities of irresponsible headline hunters and trouble makers, (2) to prevent the development of public demand for drastic and impractical air pollution and smoke control legislation, and (3) to educate the public as to the difficulty of eliminating and controlling air pollution and what the chemical industry is doing about it in order to gain member companies the time necessary to solve their problems in the most practical manner." Science historian Joe Conley aptly summarizes the MCA's justification for the program:

> The new program would have both "positive" and "defensive" functions. On the positive side, it would tell the industry's story by "fostering adequate public appreciation of the industry's contributions to the health, employment, income, standard of living, and general wellbeing of the public." On the defensive side, the program would be directed at "attacking the misconceptions that tend to undermine the standing of the industry in the public mind."

The program would emphasize how the chemical industry protected the public "in matters of defense, health, and the use of natural resources" as well as what kinds of "economic conditions" were required for the industry to continue providing these protections to the public.[13]

By the late 1950s, the MCA's PR program had expanded into a broad range of activities. Emulating the initiatives pioneered by the CF&I for rail and AISI for steel, the MCA created two publications, a *Chemical Industry Facts Book*, distributing around 300,000 copies to groups including media, banks and investment houses, schools, and members of Congress; and *Chemical News*, of which around 32,000 copies were distributed to editors, government officials, educators, and other "opinion leaders." This was supplemented by a centralized Information Service and a dedicated program of community relations.

Still, the MCA was uneasy. Retired US Army General John E. Hull, president of the MCA from 1955 to 1963, expressed this uneasiness at a speech at the National Conference on Air Pollution, organized by the Public Health Service of the US Department of Health, Education, and Welfare, in November, 1958:

> The area of scientific investigation offers probably the best hope for sensible and effective control of air pollution. But it ties in with another area of equal importance—that of public understanding. In my opinion this is the part of the problem where we have been the least successful, and I think all of us share the blame. There is ample evidence to show that the layman, the citizen not acquainted with the technology, has a fear of air pollution out of all proportion of facts as they exist. Our problem of air pollution cannot be sensibly solved without honest, accurate public understanding, and this

imposes a very important responsibility on all of us connected with the problem. . . . [T]he march of American history has proved time and time again that an honestly informed citizenry is the best guarantee of a solution to such a national problem. I earnestly hope that everyone at this confer- ence, and anyone interested in the progress of air-pollution control, will re- member this obligation and regard it as a sacred trust.[14]

Hull misdiagnosed the problem of public understanding; or rather, he framed it as a problem for which his audience at the conference had the solu- tion. But it was not that publics didn't *understand* the role of chemical com- panies and their peers in industrial power; it was that they were becoming aware of precisely how much power these industries wielded and how this power was contributing to the pollution that darkened their skies, dirtied their water, and caused mysterious illness in their populations. It seemed the price to pay for progress was becoming too high. Hull wanted a version of public understanding that restored legitimacy to the chemical industry. This was the task to which PR, as a technology of legitimacy, would devote itself.

Silent Spring: Hearing and Seeing the Environment

In 1957, as the MCA was shoring up its public information programs on pollution, a young news editor from Alabama named E. Bruce Harrison moved to Washington to work as a press secretary for Congressman Kenneth A. Roberts. Roberts, Alabama's Democratic representative on Capitol Hill, was centrally involved in the development of health and safety bills for the protection of citizens. During his time in Congress he headed federal subcommittees on consumer safety, traffic safety, and public health, and sponsored legislation to advance these causes. Harrison's role as press sec- retary included collecting the latest research and information on pollution, hazardous chemicals, and other threats to public health. Harrison became fa- miliar with members of the MCA, who regularly met with the congressman to offer their views on these legislative efforts. The MCA's perspective on public health appealed to Harrison more than did the perspective of his con- gressman: four years later, in September 1961, Harrison left his job as press secretary to take a staff position on the MCA's public relations committee.

Harrison's initial tasks involved broadening the trade association's relationships with local communities. Starting in 1954, MCA had been running an annual event called Chemical Progress Week (modeled on Oil Progress Week, sponsored by the American Petroleum Institute). Its role was to "emphasize the contributions of chemistry to individuals in their communities" by encouraging trade association members to give speeches at schools, chambers of commerce, and women's clubs; appear in radio and TV interviews; prepare exhibits in storefronts and hotel lobbies; and promote their companies and the industry at large via advertising campaigns.[15] Harrison would build on efforts to expand the positive public perceptions generated by Chemical Progress Week into a year-round endeavor, overseeing local Chemical Industry Councils whose members would "foster, through responsible inter-relationships with neighbor communities, an environment of public acceptance and goodwill in which the chemical industry can continue to function profitably."[16]

Nine months after Harrison arrived at MCA, his role as public relations manager would take on dramatically different proportions. The publication of *Silent Spring* by Rachel Carson would overturn the association's efforts to create an environment of public acceptance or goodwill. It would also inadvertently launch Harrison's career as a communication strategist dedicated to countering the rise, consolidation, and political power of the environmental movement in the United States.

Rachel Carson was already an established science writer when *Silent Spring* appeared. Trained as a marine biologist, then editor-in-chief of the US Fish and Wildlife Service, Carson had authored two bestselling books about the natural world in the 1950s, *The Sea Around Us* in 1951 (which won the National Book Award and sold over a million copies) and *The Edge of the Sea* in 1955. *Silent Spring* was of a different order. From its first pages we are apprised of the ecological catastrophe wrought by the use of lethal pesticides. Carson's poetic narrative only reinforced the dramatic scale of environmental contamination, as she contrasts "the impetuous and heedless pace of man" to "the deliberate pace of nature," quietly yet powerfully condemning humans' race to produce, pollute, and profit. Her primary target was the chemical industry and their wanton disregard for the toxic effects of their synthetic formulas in bodies and in the natural environment. The book is dedicated to the Nobel prize-winning theologian and medical missionary Albert Schweitzer. Its epigraph, quoting him, makes the book's theme clear: "Man has lost the capacity to foresee and to forestall. He will end by destroying the earth."

It would take public relations counselors at least ten years to formulate a strategy of response. In the meantime, the decade from 1962 through 1972 only reinforced the divide between "us" (citizens dedicated to reforms in environment, health, and safety) and "them" (industrial players destroying the natural environment). As the notion of the external environment came to connote public welfare, biological and ecological knowledge, and nature in urgent need of restoration, so "industry" became associated with danger, deviousness, and the hard limits of progress. The declaration of purpose of the Clean Air Act passed in 1963—introduced by Alabama congressman Roberts along with Senators Abraham A. Ribicoff of Connecticut and Edmund S. Muskie of Maine—articulated this divide in no uncertain terms:

> The growth in the amount and complexity of air pollution brought about by urbanization, industrial development, and the increasing use of motor vehicles has resulted in mounting dangers to the public health and welfare, including injury to agricultural crops and livestock, damage to and the deterioration of property, and hazards to air and ground transportation.

This new concept of the environment, especially in its antagonistic position toward industry, sent public relations efforts into a tailspin. *Public Opinion Quarterly* polls showed that industry was "most often blamed for air and water pollution and that the chemical and oil industries are the industries most often blamed." Electric power and automobile companies were also held responsible. Meanwhile, these polls revealed that "70 percent of survey respondents do not know what industry is doing to fight pollution."[17] These polls, and the concept of public opinion they harbored, were themselves undergoing a transformation in legitimacy. After 1946, fledging and discrete initiatives in public opinion research, mainly in military and applied social research contexts, were united and professionalized with the founding of the American Association for Public Opinion Research. The 1950s and 1960s saw a massive expansion of polling research by academic institutions, commercial markets, federal agencies, and political candidates, which in turn boosted the perception of public opinion as important and, significantly in this instance, as less influenced by the monolithic narratives of industrial accomplishments.[18]

Like its antecedent, the Air Pollution Control Act of 1955, the Clean Air Act of 1963 had also earmarked funds for scientific research on air pollution. But just as the belief in industry-sponsored narratives was on the wane, so too was the legitimacy of industry-sponsored research. Since the Second World War, increased government funding for basic scientific research, changing ideas about scientific evidence and risk, and growing scientific attention to problems deemed "environmental" had begun to pry open the tightly closed network of corporate-led science and its institutional and financial supports. Increased federal funding for scientific research had not only expanded the kinds of research undertaken; it had brought research on pollutants and toxins out of industrial laboratories and onto more neutral territory. New research, circulating in more public venues, reconceived industry's "threshold studies" (i.e., pollutants are permissible up to a certain threshold of tolerance) and embraced predictive science, demonstrating that even small amounts of toxins could have long-term negative effects.

In sum, industry lost its authoritative hold on the ability to decide what was and wasn't good for the public. And corporate public relations was no longer able to define, much less manage, the "external environment." Unlike the crises faced in the interwar period by coal, rail, and steel producers, the environmentalism that arose in the 1960s could not be factored into the price of progress, "informated" out of existence by new renditions of the facts, or subsumed in the spirit of political collaboration. A different relationship among information, environment and publics was needed if companies were to reckon with a newly conscious and determined environmental public. This relationship would take shape in a reformed infrastructure of public relations and the reframing of contentious industries as vital sources of energy.

E. Bruce Harrison and the Invention of Green PR

Just as Rachel Carson's work is often heralded as the point of origin and motive force of the environmental movement, so might we be tempted to see the origins of its countervailing action in the efforts of a single individual. In the sheer breadth of his attempts to produce the environment as a problem that public relations could solve, in the cleverness of his multivalent and long-term strategies on behalf of contentious industries, and in his charismatic, effusive, and self-effacing style, E. Bruce Harrison is a master architect of what has come to be known as "corporate environmentalism."

Over his forty-year career, Harrison would develop strategic information and communications programs for hundreds of companies in virtually every industrial sector touched by environmental regulations and the threat of restrictive legislation. These programs were designed to allow his industrial clients to refit, overcome, or sidestep restrictions on environmental pollution at local, state, and federal levels. His ability to remain relatively unknown by the publics he worked so hard to influence is attributable to his skill as a communicator and his ingenuity in creating opportunities for public relations where none had previously existed. Harrison played a key role in expanding the purview of "environmental" public relations for companies and in directing its subsequent impact on the concept of American environmentalism.

Harrison had left the Manufacturing Chemists' Association in 1967 for a position as head of public affairs with the Freeport Sulfur Mining Company.[19] His foray into "mining capitalism," as Stuart Kirsch terms it, took him to New Guinea and Indonesia to promote Freeport's gold and copper projects. Still, Harrison retained his position on the public relations committee of the MCA.[20] This would prove useful when he returned to Washington in 1972 and learned that his colleagues at the MCA, the American Petroleum Institute, and other trade groups and companies in major polluting sectors were struggling on several counts.

New bodies of scientific evidence detailing the scale and scope of environmental problems had prodded federal bodies to heed their results. The passage of the Water Quality Act (1965), the Clean Waters Restoration Act (1966), the National Environmental Policy Act (1969), and the Occupational Safety and Health Act (1970) as well as the tightening of existing provisions in the Clean Air Act (1970) and the Clean Water Act (1972) were bolstered by the establishment of the Environmental Protection Agency (EPA) in 1970. Nongovernmental environmental organizations also sprouted or flowered. John Muir's legacy organization, the Sierra Club, tripled its membership between 1965 and 1975.

Through his network of associations with "folks on the Hill," Harrison discovered that industry-friendly members of Congress and some major labor unions were also unhappy with the outcomes of hearings on environmental legislation.[21] To Harrison, it became evident that multiple groups could use some collective representation. And he was ready for a new challenge.

It was on this basis that the National Environmental Development Association (NEDA) was formed in 1972. Drawing together colleagues from the chemical, petroleum, and mining industries; market-minded members and former members of Congress; labor groups; and agricultural interests, all with a bone to pick over the restrictions of environmental standards, Harrison opened his own public relations agency, Harrison & Associates, and set up NEDA as his first client, with himself in the role of NEDA's executive director.[22] Companies paid subscription fees to join, and they benefited from cross-sectoral representation and the strength of a unified voice on environmental affairs. (Appendix 2 shows a full list of E. Bruce Harrison clients, including all NEDA members.)

At one level, NEDA operated according to the logics of information management and communication strategy Harrison had learned during his work with the chemical trade group. The NEDA executive committee carefully monitored pending environmental legislation, preparing in-depth analyses of recent research and polls about the issues as they played out in Washington. Regular newsletters, issues workbooks, briefings, guides to legislation, fact sheets, and printed reports all allowed NEDA members to communicate about the issues to employees, investors, and state and local representatives in the communities where they operated. NEDA also created contacts at media organizations, delivering information with the industry viewpoint on environmental issues to media outlets and specialized trade publications.[23]

The group also played a lobbying role, with members of the NEDA executive committee testifying before Congress to call for reduced constraints on environmental rules in the name of economic growth.[24] In 1973, hearings took place to review the implementation of various provisions of the Clean Air Act amendments of 1970 before the Subcommittee on Public Health and Environment of the Committee on Interstate and Foreign Commerce. Thomas A. Young, in his new capacity as president of NEDA, provided this overview of NEDA's public face:

> National Environmental Development Association (NEDA) is a non-profit, non-political, non-stock corporation comprised of labor, agriculture, industry and other private and public interest organizations and individuals. NEDA was established for the purpose of promoting the conservation, development, and use of America's resources to enhance "the quality of its human environment" (42 USC 4332). NEDA endeavors to do this by encouraging public awareness and informed input on such proposed or prevailing public policies as may serve to attain or impair that overriding human goal.[25]

As with the vague objectives of "public awareness" and "informed input," the name of the organization deliberately obscured its political stance. It was clear, in the early 1970s, that there could be no such thing as an "anti-environmental" organization. Instead, NEDA tracked the social values of the era, using its name to assert its environmental commitment while finding discreet ways to insert industry viewpoints. In one relatively unsuccessful example of this strategy at work, *Washington Post* journalists researching the sources of support for a bill to limit pesticide control introduced by the American Farm Bureau Federation and the National Agricultural Chemical Association commented:

> We were puzzled for awhile about the lobbying effort of the National Environmental Development Association on behalf of the Poage-Wampler measure. But we have now discovered that the association is run by such dubious environmentalists as Ashland Oil, Pacific Gas & Electric, and the big fruit and vegetable growers.[26]

NEDA and the Campaign for Balance

The leitmotif of NEDA was "balance." Initially, balance meant an equal consideration of economic growth alongside environmental protection.[27] To this end, NEDA would join other public relations outfits to develop what one PR professor called "a new kind of economic education program aimed at the more educated classes of the American public."[28] This public education program would focus on the "trade-offs" faced by publics in their desire for environmental protection. That is, if Americans were committed to clean air, water, and land, they would have to accept that this might come at the cost of other benefits: employment, rising GDP, and internationally competitive systems.

This view was strongly advocated in the *Public Relations Journal*'s May 1973 issue, its first ever to focus on the environment. In these new challenging conditions, public relations had a mammoth task ahead to promote the value of business. As the issue's opening editorial noted:

> The environment will be a continuing arena for clashing advocacy. No new enterprise of significance will advent without some public cost-benefit exploration of its ecological impact. The low profile will have little place as disclosure, compliance and enforcement legislation, and a more aggressive public opinion, compel open discussion of plans and problems.... [I]f

ever an area of concern cried out for the basic principles of public relations practice—keeping well informed, anticipating, recommending timely action, communicating with truth, clarity and completeness—the environment is *it*.[29]

One article in the issue, titled "Communicators and their Environmental Problems," put a point on the difficulty PR people might face in trying to reinforce the idea of trade-offs:

> Many challenges to improve the "quality of life" are not mutually compatible. In fact, they may be contradictory—in head-on conflict—or they may be capable of compromise. We must decide, and help our principals and the public decide, whether we are ready to trade off some of the presumed risks for needed benefits.[30]

Other articles pressed ideas of the "human environment" and "externalities" into service, trying to show how economic and technological issues could be counted as priorities equal to those of protecting the natural world.[31] PR needed to emphasize "rational environmental problem-solving" by insisting to publics, media, and government that ecological consequences by definition included economic considerations.[32] If one objective here was to offset the "us versus them" instilled by *Silent Spring* and its aftereffects, a second and more lasting aim was to influence the information landscape on which decisions could be made. In these "calls to action," public relations actors created and shaped categories of information as necessary components of legitimate democratic debate, hoping also to strengthen their role as brokers of such debate.

This rhetorical strategy would prove difficult to sustain. The energy shortage in the early 1970s slammed a brake on Americans' rapidly improving style of living in the postwar moment. After decades of prosperity, middle-class Americans were confronted with a sudden awareness of its potential to end. A year earlier, the publication of the Club of Rome's *The Limits to Growth* as well as the proliferation of groups such as Stanford entomologist Paul Ehrlich's Zero Population Growth had begun to instill a condition of what Lawrence Buell has called "depletion anxiety."[33] The ethic of environmental conservation, stewardship, and thrift was more in line with this anxiety than was the rationalization offered by trade-offs and cost-benefit schemes.[34]

Paradoxically, a different opportunity to promote "balance" would present itself via the energy crisis. In 1973, the Organization of Arab Petroleum Exporting Countries leveled an oil embargo on the United States in retaliation for American support of Israel during the Yom Kippur War. The drastic cut in supply led to a jump in oil prices by 350%. The shortage led to a new kind of visibility for oil and for American politics and economy, as the downturn revealed the country's fragile hold on domestic production in the context of geopolitical strife.

As Timothy Mitchell has shown, however, the energy crisis was not really a crisis at all. The production of a series of industry-generated facts and figures, mechanisms of collusion, and the projection of scarcity allowed the market to take pride of place over the state and over foreign relations.[35] Crucially, it also allowed for the creation of a new field of "energy" as a domestic assemblage of industrial power sectors, resource extraction, and policymaking. In political speeches leading up to and in the immediate aftermath of the shortage, energy gained traction as a new object of scarcity, one that required careful nurturing in order to sustain.

By producing energy as a "rival object" of scarcity to the natural environment, political and economic leaders shifted depletion anxiety away from the finite resources of the natural world and toward the finite resources of hydrocarbons and other forms of fuel. Indeed, "the politics of energy was simultaneously a politics of the environment," linking the two in a zero-sum game.[36]

The production of energy as a system in need of redress joined together a wide variety of industries affected by environmental regulations. The process offered a new set of coordinates for public relations agents to work with. In 1974, the Public Relations Society of America (PRSA), the profession's national association, organized an "energy briefing" with cabinet members at the White House. The purpose of the event was in part a reflexive exercise in legitimacy making, providing "energy-oriented PRSA practitioners with an opportunity to meet top energy officials in the same type of high-level briefing afforded chief executive officers."[37] But it would also help cement the various components of this new rival object, energy, in public communication.

Slowly, projects of questionable environmental benefit moved forward in the name of Americans' need for energy. The Trans-Alaska pipeline project is the best-known example. The project was intended to develop oil reserves and carry oil from the Alaskan North Slope to the US mainland. In 1970, a year after a series of oil spills had brought the industry's so-called externalities back into public view, environmental and indigenous groups won a federal injunction against the project. The injunction provided yet another kind of visibility, this time of the strength of environmental groups in protecting both indigenous territories and wildlife in the region.

Dedicated public relations efforts by oil companies explicitly joined energy to domestic economic independence, promoting these as part of the American individualist and entrepreneurial spirit. Mobil Oil took considerable credit for the reopening of the pipeline project, citing its advertising campaigns and other media relations efforts as a key cause. As the company stated in its internal report on public affairs over the decade from 1970 to 1980, "[In our ads] we addressed the need for continued economic growth, for which more energy would be needed, as the only way to provide higher living standards for poor people, both in the U.S. and around the world. We believed this emphasis was extremely important in a decade when thought patterns were unduly influenced by the Club of Rome's *Limits to Growth*-type thinking."[38]

Nuclear energy production was similarly promoted. The rise of antinuclear activism was partly related to the proliferation of environmental groups that recognized nuclear power as a major threat to human life. The emergence of a pro-nuclear movement (primarily made up of industrial interests) in the early 1970s was in large measure an outcropping of the new field of energy. A key strategy of the pro-nuclear movement was to expand its focus from merely promoting nuclear energy to the "promotion of other forms of energy (e.g., coal), attainment of economic growth, defense of the 'American way of life,' support of a free-enterprise economy, and independence from foreign oil."[39] As sociologists Bert Useem and Mayer Zald explain, the widened focus allowed industry groups to gain further legitimacy in two ways: first, by expanding their base of support to include people of color and women, who saw nuclear power as a potential opportunity for social and economic mobility; second, by yoking energy to national values as a response to anti-nuclear sentiment. An electric utility company manager overseeing the building of a nuclear power plant expressed this perspective in the pages of the *Public Relations Journal*:

The issue of nuclear power really is a fight over different economic, social and political philosophies. . . . We should point out that energy shortages mean cold, darkness, and great inconvenience, that the consequences of no nuclear power, or delayed nuclear power, are loss of jobs, paychecks and reduced opportunities available to our children. . . . If this point can be made in a clear way that strikes an emotional chord with the public, the activists, perhaps, will not be seen so much as being against nuclear power or the electric utility companies, but as being against the people themselves.[40]

The Three "E"s: Objects of Politics

These examples of coordinated promotion across contentious industries may appear to stretch the credulity of those doing the promoting. As Hannah Arendt has noted, "organized lying" contains within it a core of self-deception in addition to the deception of others. The self-consciousness of public relations actors relative to the strategic narratives they produce is one feature of the landscape on which environmental politics plays out. But the larger aim here is to reveal to what extent public relations relies on transforming not merely the messages but also the contexts in which these messages are understood and acted on by their publics. When PR practitioners promote energy as necessary for higher standards of living by everyday Americans, they invoke powerful relationships between individuals and their social and material environments, shaping both patterns of thought and strategies of action. To make the abstract idea of energy tangible and material to its publics, PR practitioners rendered it legitimate as part of a triad, joining energy to environment and economy to create a new network of factuality.[41]

Harrison and his team adopted this systemic understanding of energy and propped it up as the third pillar of "balance," along with environment and economy, in their call for lighter regulation. The 1976 National Conference on EEE (Environment, Economy, and Energy) Issues, sponsored by NEDA, brought together government administrators, members of Congress, and university researchers to provide perspectives on what it called a new "environmental ethic" that recognized the role of energy and economics in the calculus of the good life.[42] Riley S. Miles, executive director of the Water Users Association of Florida, a citizens' lobby group advocating economic considerations of Florida's freshwater resources, chaired the conference on NEDA's behalf. Miles made the linkage among environment, energy, and economy explicit, telling the 150 attendees that "the 'E's' are bound together in a complicated, cross-affecting manner. They cannot be separated. They cannot be dealt with separately."[43]

Harrison's efforts to promote the "three Es" were directed at not only public and political audiences but at the public relations industry as well. In 1976, with Harrison at the helm, the Public Relations Society of America created its first Energy/Environment Task Force.[44] He began writing a regular column, also called Energy/Environment, in the PRSA's trade journal. From 1977 through 1982, Harrison offered briefings, interviews with federal and other government agency officials, and opinion pieces about the intersection of energy, economy, and environment in most issues of the journal as well as in industry trade journals with interests in the outcome of environmental politics.[45]

The structure of NEDA portends a new conception of the environment as an object of politics. To understand its role in transforming the environment, we need to see it in terms of the alliances it created, the kinds of knowledge it brought to light, and the type of advocacy it embedded in the public imagination. NEDA, and Harrison himself, reoriented the role of public relations consultants as influential actors in the politics of environmental governance. In his ability to create, gather, and distribute industrial expertise; articulate core values and beliefs; and set common goals for companies around environmental concerns, Harrison claimed for his profession a determinate authority over the nature of environmental problems and the strategies required to solve them.

In terms of alliances: unlike the trade associations, or even the dominant business organizations of the era, such as the Business Roundtable and regional chambers of commerce, NEDA brought labor inside its advocacy structure. While on the surface the strategy mimicked the employee representation plans Ivy Lee helped craft for the Rockefellers in the 1910s to "bring the outside in," as we saw in chapter 2, this version was more adept at treating labor leaders as equal voices in the endeavor. One public relations trade journal observed that NEDA members "saw this as an unusual opportunity to work, at least in this narrow area, with organized labor . . . something the traditional business organizations to which the companies belonged did not offer."[46] This allowed NEDA to operate in collaboration instead of competition with the more established business groups. By 1978, the NEDA executive committee included representatives of the International Union of Operating Engineers (J. C. Turner), the Laborers' International Union (Angelo Fosco), and Associated General Contractors (Joseph P. Ashooh).

Current and former government or government agency representatives also joined NEDA. Kenneth J. Bousquet was a founder and executive vice-president of the coalition as of 1972, after a twenty-year career as a staff member of the Senate Appropriations Committee and chief counsel of its Public Works Subcommittee.[47] As a consultant to the General Atomic Corporation during this time, Bousquet could help establish connections between the chemical industry and players involved in its Cold War–era investments in atomic development.[48] In 1979 Harrison engaged a key figure, John R. Quarles Jr., whose prior role as deputy administrator of the US Environmental Protection Agency would prove enormously beneficial to NEDA. After working with the EPA from 1973 to 1977, Quarles had moved to the law firm Morgan, Lewis & Bockius. Bousquet, Quarles, and other NEDA members made use of their various affiliations, current or former, to appeal to different stakeholder groups depending on the context. The arrival of Quarles into NEDA lent considerable heft to the group. His EPA credentials gave his views on environmental protection a gravitas that his industry peers could not match. His ongoing efforts to "prune back the regulatory thicket surrounding the air pollution program" and his regular appearances on behalf of NEDA, including meetings with EPA officials, helped to attract more members. By 1981, nearly forty major national corporations as well as the seventeen unions of the Building and Construction Trades Department of the AFL-CIO were part of NEDA.[49]

A second consequential aspect of NEDA's alliance structure was its creation of internal subgroups organized around specific regulatory or legislative issues.[50] Harrison & Associates (renamed the E. Bruce Harrison Company in 1978), set up each subgroup as a fee-paying client of the firm. In 1979, the NEDA Clean Air Act Project (CAAP) was formed, followed in 1982 by the NEDA Clean Water Project (CWP), and in later years by NEDA-Ground Water (founded in 1985 to deal with legislation around toxic substances), NEDA-RCRA (Resource Conservation Recovery Act, founded in 1986 to promote industrial views on the management of hazardous wastes), and NEDA-TIEQ (Total Indoor Air Quality, working with tobacco companies). Quarles was chair of NEDA-CAAP from 1979 through 1986.[51]

As an issue-specific rather than trade-specific organization, NEDA created a new logic of collaboration to help its participants resist environmental restrictions. It could deploy various strategies of collective mobilization, rallying different members of the network for different causes. Shipping operators and contractors could come forward to argue against one piece of legislation while mining and air conditioning companies could join forces on another.

Table 3.1. Members of the National Environmental Development Association (Clean Air Act Project), 1981

Members of NEDA-CAAP in 1981	Industrial Sector
Allied Chemical Corporation	Chemicals
Ashland Oil, Inc.	Petroleum
Atlantic Richfield Company	Petroleum
Building and Construction Trades Department, AFL-CIO	Trade Union
Campbell Soup Company	Manufacturing
Celanese Corporation	Chemicals, Fibers
Chevron USA, Inc.	Petroleum
Consolidation Coal Company	Coal
Crown Zellerbach	Pulp and Paper
Dow Chemical Company	Chemicals
Dravo Corporation	Shipbuilding
E. I. du Pont de Nemours & Co.	Chemicals
Exxon Company, USA	Petroleum
Fluor Corporation	Energy and Chemicals; Mining
General Electric Company	Manufacturing
General Motors Corporation	Automotive
Getty Oil Company	Petroleum
International Paper Company	Pulp and Paper
Kaiser Aluminum & Chemical Corporation	Chemicals, Aluminum
Mobil Oil Corporation	Petroleum
Occidental Petroleum Corporation	Petroleum
Pennzoil Company	Petroleum
Phillips Petroleum Company	Petroleum
PPG Industries, Inc.	Paints and Coatings
Procter & Gamble Company	Manufacturing
Shell Oil Company	Petroleum
Standard Oil Company (Indiana)	Petroleum
Standard Oil Company (Ohio)	Petroleum
Stauffer Chemical Company	Chemicals
Sun Company, Inc.	Petroleum
Tenneco Chemicals, Inc.	Chemicals
Texaco Inc.	Petroleum
Texas Oil & Gas Corporation	Petroleum
Union Oil Company of California	Petroleum
Union Pacific Corporation	Transport
Westvaco	Pulp and Paper, Chemicals
Weyerhaeuser Company	Timberland

Source: Clean Air Act & Industrial Growth: An Issues Workbook for the 97th Congress. National Environmental Development Association Clean Air Act Project, 1981.

Another purpose of issue-specific coalitions was to allow NEDA's advocacy structure to mirror the process of federal regulation. As government regulators moved into new arenas, borrowing the standards and research established in the Clean Air Act and applying these to protection of water resources, for example, so could NEDA isomorphically apply the same strategies to counter them. Standardization of knowledge and tactics became an important piece of the group's effectiveness.

NEDA's strategy unfolded in three parts. First, the coalition would concentrate resources on making itself known as "a credible, visible entity" and "a responsible source of commentary on CAA [or other legislative] problems . . . to begin demonstrating that the regulations are causing and are apt to continue causing serious adverse effects on energy use and economic development." The repertoire of tactics to this end involved both mediated and face-to-face representation. Mediated representation took the form of press releases and press kits, editorials, and source commentary in national news media, with a focus on two major newspapers, the *Wall Street Journal* and the *Washington Post* as well as specialized publications such as *Environmental Health Letter* and the Bureau of National Affairs' *Environment Reporter*. Direct mailings of "EEE" issues went to environmental officers of Fortune 500 companies, governors, chairs of state-level legislative committees dealing with environmental or energy legislation, air pollution control officials, trade associations and other business and labor groups, AFL-CIO building trades union leaders, and members of Congress. Other, more direct attempts to draw the attention of Congress to NEDA involved personal letters written to government representatives, proposed amendments to energy and environmental legislation, and appearances at relevant hearings. The creation of events, in the form of regular conferences, "brown bag" luncheons, workshops, and meetings with environmental groups ensured the persistent presence of NEDA on the political landscape.[52]

The second piece of the strategy was to concentrate the attention of the group on environmental problems that could reasonably be solved. Another way of putting this is to say, along with Conley, that "affected industries treated the new environmentalism as a political and cultural force to be strategically managed."[53] By turning the environment into a problem of efficiency, planning, marketing, and innovation, NEDA could devise managerial solutions to accommodate it, defining responses to the policies and standards for environmental protection put in place by the federal government and its agencies.

In the case of the Clean Air Act, for instance, NEDA turned its attention to those categories of concern related to air quality standards, especially in relation to the siting and construction of new industrial facilities. The 1977 amendments to the Clean Air Act largely consisted of new provisions to ensure what was called the Prevention of Significant Deterioration (PSD) of air quality by establishing National Ambient Air Quality Standards (NAAQS) and state-by-state operating programs (State Implementation Plans, or SIPs).

A "technical subcommittee" of NEDA, relying on the insider knowledge of John Quarles, worked to identify and document weak spots in the regulatory framework established by the Environmental Protection Agency. The subcommittee would then prepare "concept" or "issue" papers with suggestions for alternatives to regulation that might find favor with the EPA.[54] One of NEDA's proposals was to adjust the standards on environmental pollutants to accommodate special cases where, they argued, exceedances might occur, such as weather fluctuations; errors in computer models that forecast pollution emissions for as-yet unbuilt new plants; and certain pollutants that might create misreadings of air quality.[55] These proposals were framed as demonstrating "greater sensitivity to local needs" such as "the social benefits of new development" and opportunities to create new energy resources.

NEDA's big push was toward decentralization, from federal to state authority over environmental rules. States, NEDA claimed, "are much closer to living with the problems of improving air quality" and are therefore better placed to evaluate it as a holistic concept. Devolving powers from federal to state agencies would let NEDA have greater influence over shaping the response to environmental concerns. This initiative would get considerable forward momentum with the arrival of Ronald Reagan into office in 1981.

In the interim, NEDA worked to develop a third component of its strategy: the creation of "grassroots" constituencies who, through public demonstrations of collective action and shared values, would manifest citizen, local, and state government support for NEDA's changes to national environmental policies.

Just as Harrison had created a client by forming NEDA, so did he apply the same kind of approach in forming publics to support it. Harrison prepared a series of lists: one of states and local constituencies targeted for early "outreach" efforts, typically in highly industrialized and unionized states where likely supporters of amendments to environmental requirements could be found; one of single-interest groups whose objectives aligned with those of NEDA, such as Americans for Energy Independence; and a third list of industrial facilities operated by NEDA member companies whose employees could be incited to join the cause.

NEDA then organized a series of site visits. NEDA "teams," composed of one industry and one labor representative, would visit companies in chemical, oil, paper, and other sectors, asking them to develop their own cases of negative impact from the legislative amendments. These cases would then be used for further mobilization and for publicity efforts to company employees, residents, and local business and government groups.

These local publicity efforts were amplified by Harrison and Quarles through op-eds and magazine articles stating their case. "This provision, like a loose cannon on a pitching deck, threatens a path of destruction," wrote Quarles in an op-ed titled, "The Clean Air Amendments," for the *Wall Street Journal.*

> The use of this radical sanction reflects a desperate gamble by Congress, hoping that the threat of economic calamity will bludgeon states and localities into adopting whatever measures are needed to achieve the air quality standards.... Even if best efforts are made, this law is likely to produce unacceptable impacts in some areas. The sooner these effects are clearly identified, the better the changes may be that Congress can further modify the statute to produce needed flexibility before it is too late.[56]

While Quarles spoke to a national audience through his op-eds, Harrison focused on local groups. "Speak up," Harrison advised company managers at hydrocarbon processing plants and offices:

> The options for concerned managers involve *aiding state agencies* as they try to shape "state implementation plans" (SIPs) for consideration of the Environmental Protection Agency before December 31; *communicating with legislators* in the states and in Washington about the real impact of CAA '77, and *publicizing* generally the tough job of complying with the law and its timetables, and the consequences of failure.[57]

One major rationale for the grassroots approach was to offset parallel efforts by environmental groups. The Sierra Club had received funds from the EPA in 1977, to host "citizen workshops" to raise public awareness of air pollution problems. Harrison's initiatives were meant to counter these information sessions with information of his own design.

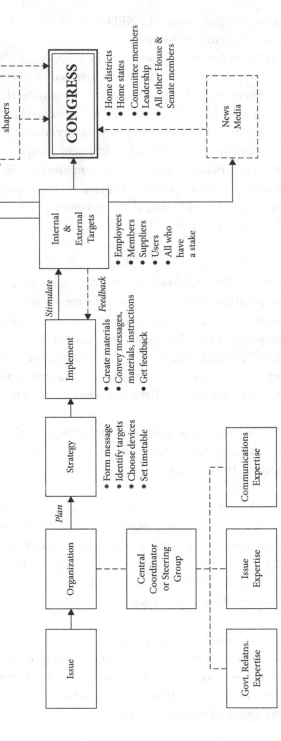

Figure 3.1. Model of Grassroots Lobbying, E. Bruce Harrison Company, Washington, DC. *Source:* Public Relations Society of America, Box 156, Folder 53, 1987.

Over the course of the 1970s and into the 1980s, the E. Bruce Harrison Company became known in its circles as the premier environmental public relations agency. In this period the firm developed dozens of coalitions to counter environmental publics, using the same template for grassroots involvement. "Grassroots involvement is the key," Harrison informed his clients. "Without it, the industry will suffer grievous damage."[58]

> To be effective today a communicator working with a public issue must indeed recognize the public-ness of the issue; he or she must reach that all-potent source of power to win on the "Washington issue": the voice and the vote of the people back home.[59]

The promotion of this new field of energy, which absorbed the economy and the environment into its ambit, created a political structure to rival the one that emerged with *Silent Spring*.[60] While both advocated a form of collective participation, equal representation, and the need for some kind of balance, Harrison's structure made the environment into a topic with which ordinary publics could not reckon. Framed as a problem of information and management, the environment became the territory of information managers. Harrison's specialized expertise would bring all manner of seekers of solutions to this kind of environmental problem to his doorstep.

4

PR for the Public Interest

The Rule of Reason and the Hazards of Environmental Consensus

On 6 October 1976, an op-ed by the public relations executive John Wiley
Hill appeared in the *New York Times*. We met Hill in chapter 2, when he co-
founded the PR firm Hill & Knowlton in 1927. In the intervening forty-nine
years Hill had grown the business from a small Cleveland concern into a
team of 560 employees in thirty-six offices in the United States and eighteen
abroad, the largest public relations outfit in the world.[1] Though Hill had retired as CEO and chairman of Hill & Knowlton in 1962, he maintained a position on the firm's policy committee, preserving both his reputation and his
commitment to the industry and his clients. He continued to appear at the
office almost daily until a few weeks before his death in 1977. The 1976 op-ed
was not quite a swan song, but it did leave readers with a sense of both his
imagined legacy and his concern for its future.

"I have lived through 21 Presidential campaigns and am now suffering
through the 22nd," the PR titan began. "I have seen 18 booms and busts in my
lifetime and five wars." Through the years he had helped his industry clients
grow more and more powerful in the political arena. But now, he argued,
business was losing its credibility, a result of its self-regard, its status as imper-
sonal behemoth, and especially, its lack of attention to the public mindset. "If
there's one thing the years have taught me it is that public opinion is the final,
all-controlling force in human society," Hill claimed. "Misled and poorly informed, it can come to false conclusions and do untold damage to business,
the economy and the nation."

Hill had reason to worry. By 1976, considerable damage to his clients
had already been done. Commercial interests, or Big Business, with
the capital letters implying an epithet, were on the back foot in 1976,
as they had been increasingly since the mid-1960s owing to a combination of factors. After World War II, America's commitment to free

A Strategic Nature. Melissa Aronczyk and Maria I. Espinoza, Oxford University Press. © Oxford University Press 2022.
DOI: 10.1093/oso/9780190055349.003.0005

enterprise and consumerism had meant that large corporations enjoyed unfettered access to capital and resources and relatively unobstructed decision-making about the shape and scope of markets. Industry's place in society was well assured. In the absence of government oversight or public pushback, however, corporate management had remained within its own orbit, largely indifferent to growing concerns over its size and power. This would change. "By 1970," writes the business historian David Vogel, "the corporation—its size, social role, political impact, and public accountability—would move from a peripheral to a central position on the nation's domestic political agenda."

The "David" pushing corporate Goliaths into this harsh spotlight was the public interest movement. The emergence and institutionalization of the public interest activist movement in the late 1960s and early 1970s represented the greatest challenge to business in the era and perhaps of the entire twentieth century. The rise of citizen groups focused on countering the political power of business in this time period was spurred by a number of factors: a more educated and expansive middle class and their changing attitudes toward political participation; the transformation of legal and political structures in the aftermath of the Watergate scandal; an increase in environmental and health disasters, brought to an increasingly national audience by the news media, especially television.

For PR men like Hill, one of the greatest threats posed by these groups was not their activism itself but the fact that they acted in the name of "the public." As we saw in chapter 3, in the decades after the end of World War II, companies were seen as important contributors to the war effort, communicating their goals as being synonymous with those of society at large. Slogans like "What's good for General Motors is good for America" or the Du Pont Company's "Better Things for Better Living . . . through Chemistry" underscored major industries' dominant self-understanding as being directly aligned with the public interest. PR professionals, as managers of publics, spent their days crafting programs and campaigns to reinforce this alignment.[2]

But by 1976, the ties were fraying. The dramatic expansion and institutionalization of public-interest groups, many organized collectively around citizen rights and the contestation of corporate governance and power, had created a new sense of the social body and of the need for diversity of voices in the political process. In the aftermath of the energy crisis and in the shadow of Watergate, the ability of business to connect self-interest with public needs was at an all-time low.

It was in the environmental realm that these new voices were especially loud. Since the publication of Carson's *Silent Spring*, the political, legal, and social opportunities for environmental advocacy had only multiplied. Civil society and government concern over environmental hazards accelerated throughout the 1960s.[3] Between 1967 and 1972, four federal environmental laws were passed and five national environmental organizations established: the National Environmental Policy Act (1969), the Occupational Safety and Health Act (1970), and important amendments to the Clean Air Act and the Clean Water Act (1972); and the creation of the Environmental Protection Agency (1970), the Environmental Defense Fund (1967), the Natural Resources Defense Council (1970), the Union for Concerned Scientists (1969), and Environmental Action (1970).[4] Twenty million people took to the streets on America's first "Earth Day" on 22 April 1970. Industry was in crisis mode, accused by all comers of ignoring the environmental impacts of its output and facing major changes to its means of production.

Hill's op-ed was a call to action. Business must learn to link "the vital elements of policies, performance and communications" in a bid for "openness, forthrightness and clarity in matters of public concern," he wrote. Its task: to wrest control of "the public interest" back from those who now operated under its banner. "Business must show, by policies and acts in the public interest and by speaking out clearly and convincingly to people, that it is worthy of their support and confidence. In my opinion, the survival of private enterprise will depend on how well this job is done."

This chapter tells the story of how business succeeded in this enterprise. Over the next ten years, business would work steadily to ensure that its policies and practices were undertaken in the name of the public interest. In the realm of environmental concerns, the task was to show that business was not only *not* part of the problem but in fact part of the solution. Rather than adapt its practices to conform to emerging environmental standards, however, business took a different tack, adapting the *meaning* of the public and the public interest to make it more aligned with its own self-interested objectives.[5] To do this, business leaders made use of the values their opponents had embodied: a commitment to pluralism in political debate; citizen participation in decisions around public policy; and transparency in the political process.

For private industry to regain its voice in political life, it needed to counter the increasingly coordinated and unified citizens' movement with a coordinated movement of its own. Organizational scholars have documented the resurgence of corporate political power during this time period.[6] Barley describes the multiple pro-business populations that made up an "institutional field for shaping public policy": business and trade associations, political action committees, public and government affairs offices, law and lobbying firms, ad hoc "astroturf" coalitions, foundations and think tanks, and public relations firms.[7] Inspired by the manifesto of Lewis F. Powell in his infamous 1971 Memorandum, the populations in this institutional field echoed the values of free enterprise through a broad range of public and political channels.[8] They amplified these values through a range of strategies, many of them borrowed from their antagonistic counterparts: grassroots organizing and coalition building, "cooperative oligopolies" formed via interlocking directorates; revolving door hiring among industry institutions; and appeals to human values and emotions.[9] With these strategies in place, business would, in this era, eventually develop a structure of social and political legitimacy that would offset the gains made by citizen movements in the courts and among the public.

Central to the coordination, coherence, and effectiveness of these strategies was the integrative and communicative work of public relations. Not just Hill & Knowlton but dozens of other PR firms joined forces in this time period to reposition their corporate clients as active participants in the pursuit of the public interest. As an epistemic community—a group of experts with recognized authority over norms, rules, and decision-making around governance issues—PR actors worked to create and structure specific kinds of knowledge around environmental action. This knowledge differed from the technical, scientific, and legal information underlying the era's calculations and calls for environmental regulation. Instead, PR actors advanced a managerial authority, producing standard-setting contexts where communication around environmental issues could take place in reasonable, rational, and disciplined forms.[10]

At the heart of the public relations principle is the effort to bring different publics toward a common understanding, whether through consensus, accommodation, or compromise. As Lee Edwards and Caroline Hodges define it, PR plays a crucial role "as a discursive force in society, shaping social and cultural values and beliefs in order to legitimize certain interests over others." PR is not merely a functional process carried out by organizations;

rather, it is "a contingent, socio-cultural activity that forms part of the communicative process by which society constructs its symbolic and material 'reality.'"[11]

Compromise is a process of building equivalencies: making ideas or things that are not alike into objects that resemble one another. One way to solidify a compromise, Boltanski and Thévenot write, "is to place objects composed of elements stemming from different worlds at the service of the common good."[12] If all parties to the debate orient their cause to the idea of the common good, even opposing views can be made to appear to act in good faith and with a disinterested or altruistic approach. This is the technology of public relations in action.

Key to this endeavor is devising language and practice that is aligned with the public in question. The authority of the PR expert resides in the ability to identify and wield compromise "objects"—a set of designations and formulations that establish points of reference for all members of the debate. "A large part of the process of working out a compromise thus consists in reaching consensus as to the adequate term, finding a formulation acceptable to all—one that 'sounds right.'"[13] Understanding how PR mediates consensus and compromise allows us to understand how business succeeded, in the 1970s and 1980s, in gaining control of the public interest. As in the Progressive era, "the public interest" was a powerful constellation of ideas about the role of the public in a democracy. Set against the heartless, crushing strength of industry's self-interest, the public interest stood for the notion that citizens could participate in the political process; that a plurality of voices was endemic to democracy; and that information should be available to all.

By the middle of the 1980s, private industry had taken these qualities associated with the public interest for itself, establishing a deep foothold in the making of public policy around environmental issues. It transformed environmental problems into problems that business could recognize and act on, making the environment more manageable for business. These reframed problems were not about restitution. On the contrary, they were about maintaining the distance between the concept of the natural environment and the role of humans in its destruction. What to do about the environment had to be turned into a matter of debate; and business had to be made into the smartest and most rational party to this debate. To understand how business regained its voice, we need to look closely at the communicative techniques and technologies employed by public relations.

"Survival in an Age of Activism"

If John Hill's op-ed on 6 October 1976 diagnosed the disease afflicting PR counselors and their corporate clients, the cause of this malady is symbolized by the front-page story in the paper that same day: "Allied Chemical Gets a Fine of $13 Million in Kepone Polluting." For years, Allied Chemical had discharged process water laced with Kepone, a DDT-related insecticide, into the Chesapeake Bay in Virginia, poisoning the waterway's fish and causing neurological problems in workers who had handled the chemical.

In the sentencing against Allied, the largest polluting penalty ever levied on a company, the judge made his reasons clear. "The environment belongs to every citizen, from the lowest to the highest," he told the assembled parties in the courtroom. "As a nation, we are dedicated to clean water. I disagree with the defendant's position that this was all done innocently. I think it was done as a business necessity, to save money. I don't think we can let commercial interests rule our lives."

The devastating effects of DDT (dichloro-diphenyl-trichloroethane, a chemical compound) were more or less completely unknown to the public until the early 1960s. In the immediate post–World War II era, DDT was celebrated as a highly effective and sophisticated means to protect citizens and crops from insect-related diseases. American and British governments produced films showing people being doused with DDT to prevent polio. It was one more demonstration of the ongoing power of industry and its technical mastery over the environment.[14]

One of the many explanations for the outgrowth and institutionalization of citizen advocacy around environmental issues in the 1960s resides in the emergence of the public interest movement. Of course, "lobbying for the people"—collective action by ordinary individuals around social or political issues—is not unique to this era.[15] In its community orientation, drive for institutional change, and push to limit corporate control, the public interest movement followed the path laid by earlier movements in the American reform tradition, such as the muckrakers of the Progressive era and the labor unions of the 1930s.[16] In the 1960s, however, public interest groups took on a distinct character. The definition of public interest in this context was the pursuit of a non-economic good that would benefit ordinary people in their everyday lives. That definition placed the public interest in direct opposition to the notion of self-interest—economically or politically motivated goals that favor elite and established groups.

The Allied Chemical fine was symbolic in another way. It illustrated a signature tactic of the environmental movement to gain attention to its cause: the use of lawsuits as a means of advocacy. The Environmental Defense Fund was incorporated in 1967 following a successful suit brought by scientists and bird watchers to stop DDT from being sprayed in Suffolk County, New York.[17] Backed by the National Audubon Society and using the publicity generated by the legal case to mobilize additional supporters and funds, the anti-DDT campaign culminated in 1972 with a ban on the pesticide by the newly formed Environmental Protection Agency.[18]

Two legal concepts in particular were cornerstones in the consolidation of lawsuits as means of citizen advocacy and the further embedding of "public interest" as a collective good against private industry. These were the concept of "standing" and the concept of "class action." Notably, both emerged from environmental concerns. The doctrine of standing refers to "who has a right to be heard in court on particular issues involving activities undertaken or regulated by public agencies."[19] Until the 1960s, standing was determined by interest, and interest referred to economic interest. Those with a right to be heard had to demonstrate their interest on the basis of economic impact. In other words, standing was for private parties and not for individual citizens with a concern for the public good.

Following two precedent-setting cases, the court reasoned that this notion of standing was too limited. Citizens have "an interest in actions that affect the nature of the environment, and . . . this interest is arguably within the zone of interests that are or should be protected by law."[20] The standing of the citizen took into account "as a basic concern the preservation of natural beauty and of national historic shrines, keeping in mind that, in our affluent society, the cost of a project is only one of several factors to be considered."[21] Going forward, "citizens will be recognized in court as advocates of a public interest, on the grounds that, as members of the public, they have been or may be injured by the actions complained of."[22] Federal judges increasingly interpreted federal statutes "to guarantee a wide variety of groups the right to participate directly in agency deliberations as well as to bring their complaints to court."[23] Class action expanded this emergent right for citizens to participate in legal and regulatory proceedings.

Closely connected to the problem of the public interest, for corporate leaders, was the problem of publicity. A key figure in the public interest movement, and arguably the motive force behind such publicity, Ralph Nader and his team of public interest lawyers had been pushing forward "citizen action as a countervailing force" against big business and irresponsible government. Born in 1934 in Winsted, Connecticut, the son of Lebanese immigrants, Nader's career helped transform the format and genre of citizen and consumer advocacy. Nader, a graduate of Harvard Law School, understood not only the power of legal action but also the power of research and grassroots networks. He created a set of organizations—the Center for Study of Responsive Law in 1969; the Corporate Accountability Research Group in 1971; and Public Citizen Inc. in 1971 (which itself spawned a volunteer-run national network of Citizen Action Groups, better known today as Public Interest Research Groups, or PIRGs)—all dedicated to exposing corporate, government, and regulator malfeasance.[24]

Beyond his legal skills, his capacity for research, and his organizational prowess, Nader was an exceptional and tireless publicist. Cross-country speaking tours, press conferences, congressional lobbying, petitions and letter-writing campaigns, small-scale advertising to solicit funding contributions ("voluntary contributions solicited through paid newspaper ads and mailings"), publication of research studies and working papers, attendance at public hearings—Nader and his "Raiders," as his staff were known, wielded the power of the media in framing and amplifying their efforts, all in the service of public reform.[25] Colleagues and like-minded organizers did the same. John Gardner, profiled in the *New Yorker* in 1973 about his reform organization Common Cause, noted the need for citizen action to be supported by an informed public. "The special interests flourish in the dark. Officials begin to respect citizen action when they discover that citizens are watching and the media are reporting what the citizens see."[26] Barry Commoner, ecologist and research scientist, who spent his career demonstrating the relationships between scientific information and citizen action, in 1963 co-founded (along with the anthropologist Margaret Mead) a national organization, the Scientists' Institute for Public Information (SIPI), which for two decades worked to ensure public participation in environmental politics.[27]

As experts in matters of publicity, PR counselors were particularly worried about these activities and even more concerned with their own weakening grip on the public narrative. Throughout the 1960s and into the '70s, they had grown increasingly uneasy about the new conjunction of environment, public, and publicity, and a style of advocacy that left them in the cold. Many public relations managers diagnosed the problem as an excess of human "feeling" around environmental and other social issues, an emotional response produced by overdramatic extremists with little regard for the facts of the matter.

The trade journals reflected this mounting concern. A 1969 opinion piece in *PR Journal*, "Survival in an Age of Activism," describes a world of growing complexity and information overload that public audiences cannot digest. Instead, they are swayed by the "human feeling" conveyed by activists. "In the arena of present 'attitude management,' not the facts but the impression people get of a situation is the real reality. What the public thinks is 'real' will probably determine the result, and not the merits or the actual conditions." The greatest problem, the author concludes, is that "communication in our society is in revolution. The standard processes whereby information and ideas seep through the populace, from the top down or horizontally, cannot compete with the visible, dramatic, easy-to-sensationalize communication that results from activism."[28]

In another *PR Journal* article, "Environment: A New PR Crisis," the director of PR firm W. R. Grace & Co. also noted the power of media "sensationalism" to influence publics around environmental issues, with troubling implications for industrial PR:

> Industrial public relations men, particularly those in heavy industries such as chemicals, steel, cement, paper and petroleum—to name a few—will come to think of the 1970s as the decade that focused on every ill, real or imaginary, foisted on man by man's own need for industrial products and by the disposal of the waste materials resulting from their manufacture and use.[29]

"Far more ink and rhetoric and videotape flowed for Earth Day than for any special day or week or month that any of us ever devised," complained the director of public relations for the Dow Chemical Company. "In that sense Earth Day must stand as a publicity triumph of the greatest magnitude.[30]

A 1971 report by Hill & Knowlton pulled no punches in condemning the activists, public interest organizations, and scientists whose growing influence and coordinated efforts were creating the problem. "Slings and Arrows, Inc.: A Report on the Activists," highlighted Nader, Commoner, and Gardner as prime movers in the influence network, "able to enlist the support of millions—and the influential thousands—by pursuing causes and abrading grievances that are real enough to bring enthusiastic support—at the nation's capital and way down home."[31] The report reviewed annual directories of environmental science and conservation groups and analyzed the backgrounds of boards of trustees and advisory council members for a range of recently established organizations: the

Environmental Defense Fund, Friends of the Earth, SIPI, and the Center for Study of Responsive Law.

"These rosters show how a few dedicated people with a little money, a lot of publicity, and an idea with great appeal can today launch what appear to be mass movements, can influence politicians, harass industry, use laws and courts and regulatory bodies, enlist popular support for their objectives, and accomplish many of their objectives," the report read. It continued:

> And in pursuing their objectives some do not hesitate to use shock tactics, preaching doomsday because man is upsetting nature's balance and destroying the environment. Deliberate exaggeration is part of their strategy and they defend it as necessary to dramatize their cause and get attention. So they picket, stage rallies and demonstrate, especially when the television cameras are turning. And of course, they write, they speak, they testify and they attend, and they disrupt meetings—endlessly, but with a dedication not matched by those whom they criticize and attack.[32]

Lists of disruptions to industrial projects (power plants, pesticide applications, auto manufacturing, trash collection) and of recent and forthcoming regulatory initiatives to further dampen industrial production were accompanied by a series of recommendations. "If anything has been shown in the last few years and in the preceding pages, it is that if business doesn't take on some of these responsibilities—someone else will. . . . [I]t is obvious that if the businessman waits to be forced into action, he may find himself forced out of the action." Ultimately, Hill & Knowlton concluded, business needs to "offer better, more sensible and feasible, more viable and more honest alternatives . . . and beat 'em at their own game—if we're not too late."[33]

Milton Wessel and the Rule of Reason

Business, and especially the public relations profession, found its answer in the ideas of Mr. Milton Wessel.

Milton Wessel was an American trial lawyer who made his career in corporate practice. From 1970 until 1978, he worked with the chemical industry, as general counsel to the Chemical Industry Institute of Toxicology (CIIT), and as special litigation counsel to the Dow Chemical Company. Wessel's primary concern was what he called socioscientific disputes. These, for Wessel, were problems that were based on complex scientific or technical formulas but that were of concern to society at large and therefore required public favor in order to be resolved.

Wessel had experienced his share of socioscientific disputes. He had represented Dow in connection with the 2,4,5-T herbicide, a highly toxic contaminant developed by Dow and used in Vietnam under the infamous name, Agent Orange. He had witnessed the increasing failure of his corporate clients to win the lawsuits brought against them. But more concerning to Wessel was that these failures appeared to be the result not of justifiable guilt but of underinformed public opinion. A number of recent events, from the Three Mile Island nuclear disaster, to the increased recognition of carcinogens in food, to the awareness of air pollution caused by coal mining, were inspiring a growing distrust by the public in the benefits of scientific and technological progress. And this distrust was manifested in the number of legal cases being brought against companies as well as the desire by the public to know how these companies affected their health and well-being.

Amid this call for more information and greater transparency by industry, Wessel worried that his clients, and the legal profession more broadly, were being dragged through the mud. Wessel felt that the due process of the court was being displaced by the emotional tenor of the court of public opinion. A central problem with public-interest affairs, Wessel argued, is that while typical courtroom proceedings emphasize a focus on justice and fair process to determine the outcome, "the public wants the focus to be on 'truth' alone." Yet 'truth,' Wessel explained, "is an uncertain and sometimes most illusory concept"[34]:

> The public does not care that the rules are carefully and properly followed, which is the primary focus of our traditional adversarial mechanisms. . . . The public has great interest in the outcomes of these disputes, which involve important "quality of life" problems. It cannot adequately evaluate those results, however, because of the enormous complexity and uncertainty which are always involved. As a result, the public will be satisfied with, and accept, the decisions in these disputes only if it has confidence in the integrity of the process by which those decisions are being reached.[35]

Wessel's solution was to develop an alternative process for debate; one that took socioscientific disputes out of the adversarial and procedural arena of the courtroom and into an environment with more room for discussion, negotiation, and compromise. In a non-confrontational, collaborative setting that brought opponents to the negotiating table, companies could engage in public-interest affairs on surer footing. Rather than being labeled opponents of the public interest, pursuing due process at the expense of moral or social truths, companies could find ways to create outcomes that demonstrated their social responsibility and commitment to society's progress.

Wessel called this alternative process the Rule of Reason. The Rule of Reason was a method of resolving disputes that involved long-range planning instead of short-term wins. It entailed a vision whereby "the leaders of science, industry and society" met to "reduc[e] confrontation and introduc[e] reason and logic into the resolution process." It sought "transparency" in the process, more sources of information as evidence, and simplification of complex scientific concepts to facilitate public understanding. "There must be a major effort by all to understand the views of any opposition and to accommodate to it whenever possible."[36] In sum, it offered a managerial instead of an adversarial approach to resolving contentious issues.

Implicit in Wessel's alternative means to resolve social and environmental disputes was a deep desire to regain credibility for his clients and for his own profession. Prominent court battles between corporations and environmental groups were furthering the conceptual gap between the rapacious self-interest of business and the collective public interest of environmentally minded citizens and scientists. In a review of his 1976 book, *The Rule of Reason: A New Approach to Corporate Litigation,* the *New York Times* quoted Wessel on the motivation of his pen: "Environmentalists have discovered the soft underbelly of the industrialists. . . . They sometimes provoke the hell out of the companies and win unsound cases as a result."[37]

The Rule of Reason was therefore a response to the damage caused to business and the law on multiple fronts. It advocated an alternative path to the seemingly cut-and-dried outcomes of courtroom battles and the indisputable evidence of scientific research in environmental disputes. By urging business leaders to fight back with appeals to reason, long-term thinking, and points of consensus, Wessel was offering a chance for business to participate in the environmental sphere on a more even footing.

This insight landed in the PR community like a bolt of lightning. Here was the answer to the problems that plagued PR counselors in the environmental arena. Public relations agents could use the Rule of Reason to reposition business as a committed participant and partner in environmental problem-solving. By appearing to extend the olive branch in contentious environmental disputes, business could take on the role of the reasonable and rational party while counterposing antagonistic response by environmentalists as unreasonable and extreme. And by framing business as operating *within* the public interest instead of against it, PR communicators could demonstrate their clients' ability to heed the power of public opinion as well as regain their own authority as managers of this public opinion. To be worthy of its name, public relations needed to take back the mantle of the public interest.

PR Expands Its Authority

Public relations counselors realized that applying the lessons of the Rule of Reason to environmental problems involved a series of maneuvers. First, they needed to establish their own authority as arbiters of reason, independently of the legal profession, for it was not only business whose reputation was suffering in the 1970s. In the aftermath of the Watergate scandal, trust in the legal profession was at an all-time low. As Wessel himself delicately observed in the preface to his book, "Public dismay at the Watergate disclosures regarding the improper conduct of so many lawyers, and the burgeoning complaints regarding the inadequacy of trial attorneys, reflect the reduced esteem in which the profession is presently held."[38]

The Watergate scandal was a point of inflection for the public relations industry as well. Increasing public scrutiny in the mid-1970s and congressional reforms distributing power among subcommittees made old-style centralized lobbying ineffective.[39] For some PR firms, the solution was to gain distance—at least in appearance—from lobbying activities. But as managerial elites began to consider a stronger role in public policymaking, business groups desired more, not less, access to Washington corridors.

Traditionally, negotiating with power brokers in Congress ("government relations") and appealing to audiences in state and local arenas to gain support for a policy position ("public relations") were discrete functions carried out by separate and not necessarily related authorities. But as *Business Week* reported in 1979, "Businessmen are quickly searching for new lobbying techniques that are better suited for gaining the favor of a more independent Congress. They recognize that public opinion has greater sway over most policymakers in the post-Watergate era, and congressmen, in particular, appear much more responsive to the demands of their constituencies and less to the wishes of party and congressional leaders."[40]

In response, companies integrated the two types of advocacy, either by assembling an in-house public affairs team or by working with external PR/public affairs firms (some staffed by former employees). Public affairs slowly gained authority as an executive function rather than merely an administrative one. By connecting government relations with public relations, and by increasing the number of PR representatives both within private sector firms and in their own PR firms across different states, the effect was to dramatically increase the channels of communication of an issue, so that constituents "back home" effectively joined Washington negotiators in lobbying around questions of public policy.[41] This allowed contentious industry players to "decentralize" their efforts, impacting municipal or state populations instead of just Capitol Hill.[42]

At the same time, prominent PR firms and companies began to employ well-connected lobbyists to operate from within their firms.[43] The job of public relations itself took on a more expansive role, adding to its standard tasks technical knowledge about environmental and health problems, regulatory knowledge about environmental policy issues, and legal knowledge about navigating trials.[44] Writing in *PR Journal* in 1977, E. Bruce Harrison encouraged his colleagues to see themselves as managers and to treat the capitol as a "management system":

> Laws are not "enacted by Congress"; they are the end product of the efforts of successful managers. Regulations are not "promulgated by" a certain agency; they are the result of successful management. News and commentary are not mere outpourings "of the media" or "of the *Washington Post*"; they are the yield of planning, motivating and regulating the tasks of persons who are writers, editors, and broadcasters.[45]

In some cases, instead of working through existing trade associations or industry groups, PR counselors would create their own organizational forms to manage specific issues—especially if those issues required urgent attention. One PR expert, Matthew M. Swetonic, described his experience working for Johnson-Manville Corporation's asbestos-health management committee in the late 1960s and early 1970s. Facing growing media scrutiny over asbestos exposure, Swetonic encouraged Johnson-Manville to form a trade association that would exclusively handle the communications aspects of this issue. In this way, Johnson-Manville would reduce its own individual media exposure and create an actor to represent the entire US industry, decreasing the

ENVIRONMENTAL, HEALTH AND SAFETY AFFAIRS
- Environmental Issues Management
- Environmental Regulatory Counsel and Representation
- Communication Support for Permitting, Superfund, RCRA, and Air Issues
- Crisis Communication Planning and Support
- Community Right-To-Know (SARA) Planning and Support
- Plant / Community Relations

NEWS MEDIA SERVICES
- Reporter and Editorial Board Briefings
- News Conferences
- Media Tours
- Press Kits and Materials
- Op-ed Articles
- Public Issues Advertising

GOVERNMENT RELATIONS SERVICES
- Legislative Monitoring and Analysis
- Lobbying Support
- Regulatory Monitoring and Analysis
- Grassroots Communication Services
- Executive Branch Liaison

MARKETING SERVICES
- Consumer Opinion Surveys
- Marketing Plans
- Marketing Media Placement

CORPORATE RELATIONS SERVICES
- Annual Reports
- Company Brochures
- Annual Stockholder Meeting Support
- Special Events
- Financial Communication

ORGANIZATIONAL MANAGEMENT SERVICES
- Association Management
- Coalition Development and Support
- Operational Audits and Counsel

EXECUTIVE PRESENTATION SERVICES
- Media Spokesperson Training
- Speech and Presentation Coaching
- Government Testimony and Witness Preparation

Figure 4.1. "Many Arms": List of services provided by the E. Bruce Harrison Company environmental public relations firm, ca. 1992.

firm's direct liability. The association, created in 1970, was called Asbestos Information Association/North America (AIA/NA), and its responsibilities went far beyond what was considered standard public relations at the time. This is how Swetonic describes it: "The Association would not just deal with the media, but would create a technical information arm to advise industry members on the appropriate ways to control asbestos exposures in the workplace; a regulatory information arm to work with government agencies on the development of reasonable workplace and environmental standards; and, in the future, a legal arm to assist the industry as whole in defending itself against liability claims."[46]

With every "arm" created, we see the further reach of managerial strategy into political and social spheres (figure 4.1). Constructing the environment as an object to be managed is the outcome of concerted and ongoing control by industry actors, constituted and coordinated in large measure by PR managers. Each arm makes the environment more stable as a concept and more difficult to shift in the public mind. What had to shift, then, was the terrain on which activism could take place. Forced to do battle with an increasingly intractable idea of the environment as a product of technical information, industrial standardization, and public mediation, activists found themselves renewing their emphasis on consumer-oriented, rather than citizen-oriented, tactics. This terrain was far more familiar to industry leaders and their PR managers, allowing them to continue to set the rules of engagement.

Displacing Scientific Evidence

A second way public relations counselors aimed to instill the Rule of Reason into environmental debate was to produce a different style of negotiation that would take the place of courtroom disputes. Here they innovated with a managerial negotiation style, one designed to reach points of consensus, agreement, and compromise rather than antagonistic opposition.[47]

This was achieved by transforming what counted as evidence in negotiations. One of the problems plaguing environmental battles, for corporate actors, was the reliance on scientific expertise. The criteria used by scientists to judge the efficiency and dangers of chemicals, and especially to determine what was necessary to safeguard the public, was heavily relied on in the court cases brought by the environmental movement. As we saw in chapter 3, until the early 1960s, industry had a stranglehold on scientific research conducted on the health effects of their products. Moreover, leading up to and immediately after the Second World War, "Americans assumed that science was good, that chemicals were necessary, that these experts could be trusted, and the side-effects of chemical use would be negligible."[48] But in the postwar years, with greater government funding and more public and congressional scrutiny of the health hazards of industrial products, arm's-length scientific work uncovered serious concerns, bolstering and extending the environmental movement's impact.[49]

Wessel argued that the burden of scientific consensus—its slow, incremental, and highly technical nature—was at odds with the need to communicate to the public about society's major environmental problems. "We no longer have the luxury of awaiting a final scientific consensus in this traditional sense. Decisions *must* be made now. There is no other alternative. We either do or do not use a chemical; we either do or do not use nuclear power. To await the final, traditional scientific consensus may mean that the barn door was closed long after the animals escaped. We must find a scientific alternative."[50]

"Forming the best possible public policy decisions in socioscientific disputes requires a very different kind of scientific consensus than that of the past," Wessel argued. To foster democratic decision-making around issues of public concern, people needed more information about scientific matters, and particularly in which areas scientists *did* find consensus around how science impacted public policy. Where there are "substantial areas of agreement, the public is entitled to have the benefit of such agreements," he wrote.[51]

Wessel's paradigm emerged from a highly publicized controversy: the trial brought by the Environmental Protection Agency with the Environmental Defense Fund against the Dow Chemical Company over the toxicity of the chemical 2,4,5-T. In 1948, the chemical was registered as a pesticide in the United States and used to manage agricultural crops and control weeds. It was little known to the public until the Vietnam War, when it was used as a defoliant known as Agent Orange. By the end of the 1960s, reports emerged that the defoliant was having severe health effects on local populations in Vietnam. News reports began covering administrative and class action suits charging that the herbicide caused birth defects and cancer, raising public alarm. In the early 1970s, the Environmental Protection Agency moved to ban the chemical, sparking further media coverage of the growing controversy.

On 8–9 March 1974, a conference was held by the Dow Chemical Company to prepare for the upcoming trial. As counsel to the company, Wessel advised his client to review the scientific evidence as carefully as possible to see where it might not hold up. "New understanding or information might suggest that more testing and research were required, or that some preconceived view of the scientific evidence should be modified."[52] Although the lawyers from EPA/EDF were not initially invited, the publicity surrounding the event eventually forced their opponents to allow them to attend.

What ensued, in Wessel's terms, was an unprecedented opportunity for dialogue between adversarial groups. By examining opposing evidence and sharing points of agreement, "it became more and more clear that many apparent scientific differences were not differences at all, or were really differences in the kinds of risks people believed worth taking—value differences, and not scientific differences."[53]

Three and a half months later, on 24 June 1974, Deputy EPA Administrator John Quarles announced that the EPA was terminating its proceedings. For Wessel, it was a moment of transformation:

> Whatever future scientific research and investigation might suggest, EPA's "public-policy" decision on 24 June 1974 was that the benefits of permitting continued use of 2,4,5-T outbalanced the hazards. People might differ with this value judgement; no one differed sufficiently with the scientific evaluation to complain legally. As the result of the "rule of reason" conference, "science" had thus been factored into "public policy" with enough credibility to at least end the legal fight for the time being.[54]

In years to come, the battle would continue to rage, in and out of the courtroom. But in this moment, the outcome of the 2,4,5-T debate was to shift the basis of knowledge from scientific to dialogical norms of consensus. With a focus on dialogue, values, cooperation, and compromise, industry and its representatives could gain a stronger foothold in the policy debates.

By the 1980s, the idea of "alternative dispute resolution" (ADR) had made considerable inroads into business strategy. By 1985, at least 113 companies had signed a Corporate Policy Statement on Alternative Dispute Resolution, a voluntary pledge that commits the signatories to engage in ADR "as a method of first resort," before turning to the courts.[55] The chemical industry in particular was a staunch advocate of ADR. The president of the Chemical Manufacturers Association (CMA) sent a letter to its members encouraging them to sign the pledge. "If our industry is seen as generally inclined to consider ADR in inter-corporate disputes, that reputation may have a spillover effect when we deal with Washington issues in convincing people that we are serious about trying to cooperatively solve problems in that arena as well," the president of the CMA told *Chemical Week*.[56] The spirit of compromise embedded in ADR made it a superior strategy for polluting industries. It was hard to fault a company that embraced dialogue, reason, and joint efforts to reach agreement. But the greater effect of zeroing in on common cause and shared values was to side-step incontrovertible scientific evidence of environmental problems. It was far easier for companies to regain legitimacy through a democratically inflected commitment to dialogue and collective participation than to push over the competent and critical integrity of scientific findings.

The EPA also developed a regulatory negotiation project, including groups like the CMA, the National Agricultural Chemicals Association and the US Dept of Agriculture to work out issues surrounding pesticides.[57]

PR and the Court of Public Opinion

A third maneuver undertaken by public relations counselors from the late 1960s through the 1980s was to anoint the court of public opinion, rather than the court of law, to render final judgment on environmental issues. To a certain extent, this had already been done for them. As Michael Schudson has written, this was the era of "the right to know," in which public audiences called for increased transparency and availability of information as the "currency" of democracy. "Information and its availability to the public at large became a theme for a wide variety of reforms and reformers in just the years that Nader came to national influence in the mid-1960s and into the 1970s."[58]

The passage of environmental laws in this era, such as the 1969 National Environmental Policy Act and its "most potent element," the environmental impact statement (EIS), mandated the disclosure of the potential environ-mental hazards of any proposed legislation or other major federal action. The EIS was by decree a document subject to public review. It instilled a mech-anism of accountability, via information, into the federal government in the realm of environmental protection. Most important for our purposes, it moved the idea of the public from a beneficiary of environmental action to an active participant in its outcome and preservation.[59] It would transform both citizen interventions and institutional culture for decades to come.

As Stephanie LeMenager has pointed out, "Transitions between mass media platforms have coincided with innovations in environmental action and even philosophy."[60] The rise of television as a medium of environmental action created immediate and visible forms of evidence for environmentalists' cause. The environmental organization Greenpeace had burst onto the scene in 1970 using television as a central tactic in its direct-action protests. Greenpeace helped to create an international, middle-class audience for environmental issues, using both their own footage and news coverage of their actions to build resonance with this new public.[61]

In the national context, the task for corporate PR counselors was to find ways to create and circulate information about the environment on behalf of their clients that could match the power of the information emanating from the media and from executive agencies. It had to resonate with the values of the era: transparency, accountability, and democracy. And it had to operate in the name of the public interest.

PR representatives had already worked hard to generate a wealth of *internal* information for their clients about air and water pollution. The American Petroleum Institute, for instance, had developed extensive information banks for its members, including newsletters, bibliographies, and background papers. It prepared briefings and testimony for public hearings and sponsored research at government facilities. This internal information now needed public forums. PR agents wanted to get this information from the hands of members into the hands of the public, so it would resonate with the values of the era's "right to know." Increasingly over the course of the 1970s, PR agents developed more sophisticated relationships with media makers and opinion leaders to achieve this goal.

Advertorials—a portmanteau for advertising and editorial commentary— were one way industrial organizations aimed to insert their voice into public interest dialogue in the 1970s. Advertorials to promote a political position, also known as advocacy advertising, had been used to great effect throughout the twentieth century. Brown and Waltzer describe one of the earliest advertorial campaigns in the 1908 American Telephone and Telegraph Company push for a monopoly national telephone system.[62] The impact of industry advertorials on political discourse was dramatically expanded on 26 September 1970, when the *New York Times* created an op-ed page, assigning the lower right quadrant of the page to non-commercial speakers.[63] Less than a month later, the Mobil oil company ran its first op-ed ad.[64]

"For a free society to survive, the public must have access to the widest spectrum of news, facts, and opinions," Mobil's head of public affairs Herbert Schmertz opined in an interview a few years later. Schmertz, a former lawyer and political consultant who ran Mobil's public relations from 1970 to 1988, was the architect of the advertorial approach. "In 1970 it was our view that business in general, and the oil companies in particular, was failing in its obligation to inform the public." In addition to offering solutions to the energy shortage, Schmertz highlighted another prime function of the advertorials: "We felt that litigation, legislation, and regulation were creating problems for our nation by impeding energy production and by raising energy costs." The advertorials, therefore, were another source of non-scientific evidence in the name of the public interest. "Mobil sought to foster a dialogue by expanding the spectrum of views, opinions and facts and by alerting people to the dangers that threatened the economic health of the nation."[65] But the traditional effort of speaking to journalists was of limited use, given the tendency of the media, in Schmertz's terms, toward "simplification and distortion." By ensuring Mobil's own voice was heard, in its own words, advocacy advertising could gain the ear of the public for the problems as the company wanted them articulated.[66] Throughout the 1970s, Mobil maintained a massive advertising and op-ed campaign "with continuing emphasis on the need for growth in energy and the economy."[67]

By 1973, Mobil was placing its advertorials in five other major newspapers: the *Los Angeles Times,* the *Chicago Tribune,* the *Boston Globe,* the *Wall Street Journal,* and the *Washington Post.*[68] A year later, in an effort to reach beyond urban publics to the "heartland-community readers," they added a magazine campaign, placing its advertorials in popular magazines like *Reader's Digest, Time, Parade,* and *Family Weekly,* and in service-club magazines such as Rotarian, Kiwanian, Moose, and Elks.[69] The advertorials ran every Thursday for nearly thirty years, from 1972 through 2000.[70] As an internal report of Mobil's public affairs campaigns concluded, "In a relaxed way, these columns got across Mobil's major themes, not only the need for energy but the need for less regulation."[71]

Mobil's public relations and advocacy program extended far beyond newspaper advertorials. To respond to the television coverage of environmental action protests and what the company perceived as unfair reporting on the energy crisis and the Arab embargo, Mobil turned this medium to its advantage. One strategy involved a campaign of media "blitzes" by Mobil executives (coordinated by public relations managers) around specific issues. Between 1975 and 1977, Mobil conducted three blitzes, on Mobil's proposals for a National Energy Plan, on the question of oil company divestiture, and on the topic, "Is America running out of oil and gas?" On this latter topic, "23 senior Mobil managers . . . visited 29 cities in 21 states, calling on 30 newspapers and appearing on 69 television shows and on 68 radio programs."[72]

In addition to media relations, from 1975 through 1981 Mobil produced public service announcements for television (table 4.2). This move allowed them to sidestep television network rules limiting airtime for commercial viewpoints on political issues. These sixty-second spots, which aired regularly on around 175 stations, used third-party commentators and dealt with such issues as "offshore drilling, federal lands, and environmental protection." The company also created "news clips" for TV stations—commercials promoting Mobil's take on energy issues—that reached broad audiences (table 4.1).[73]

Perhaps the most impactful of Mobil's onscreen public relations efforts to align their company with the public interest lay in their sponsorship of public television programming. Schmertz described the purpose of Mobil's sponsorship of cultural programs in two ways. First, it was an opportunity to position the company's leaders "as corporate statesmen whose concerns go beyond the bottom line . . . and [who are] intellectually entitled to be listened to on vital public-policy issues."[74] Second, "we now find that when we give certain publics a reason to identify with the projects and causes that we have chosen to support, they will translate that identification into a preference for doing business with us."[75]

The American Public Broadcasting Service (PBS) was an ideal venue in Mobil's eyes. As public, non-commercial television, it did not carry advertising. Mobil's voice in this context was therefore perceived as public and non-commercial. Starting in 1971, its sponsorship of the immensely popular television show, *Masterpiece Theatre* (replacing the Ford Foundation as the largest sponsor of PBS) aligned the company with the genteel elitism of the English drama. Mobil (led by Schmertz) not only sponsored the show; it also selected the theme music and the host. Mobil also controlled all of the publicity for the show, presenting it as *Mobil Masterpiece Theatre*. Though the

Table 4.1. Mobil Oil Issue-Oriented TV Programs, 1976–1981

Year	Subject	Time (Minutes)	Stations	Est. Audience
1976	Divestiture	4	76	4,600,000
	Divestiture	30	21	750,000
	Gasoline Prices News Clip	2	67	2,800,000
	Solar Energy	30	80	2,700,000
	Offshore Drilling News Clip (Massad)	4	56	3,700,000
	Offshore Drilling News Clip (Clewell)	3	94	5,200,000
1977	Gasoline Prices News Clip	3	46	2,300,000
	Solar Energy	4	62	2,600,000
	Price at the Pump	30	27	1,400,000
1978	Coal	30	94	3,800,000
	Coal News and Talk	4	101	4,300,000
	Energy Dilemma – Cable	30		500,000
	Search for Oil & Gas	30	76	1,800,000
	Oil & Gas News and Talk	4	70	3,200,000
	Supply News and Talk	4	29	1,400,000
	Methanol News Clip	3	114	5,900,000
	Regulated America	30	104	2,700,000
	Regulated America (5-part)	3	74	17,000,000
	R. Warner Reaction to Energy Bill	2	58	3,400,000
	Heating Oil	2	64	3,000,000
1979	Prices & Profits (5-part)	2	103	36,000,000
	Nuclear	30	110	3,500,000
	Clean Air	30	65	1,500,000
	Capital Formation	30	102	2,400,000
1980	Energy (5-part)	2	109	5,000,000
	Energy at Crossroads	30	64	1,500,000
	Gambling on Energy	30	70	1,600,000
1981	Gambling on Energy (3-part)	2	112	13,000,000
	Oil Hunter	5	34	1,000,000
	Gas Prices (3-part)	2	123	17,000,000
	Energy Quiz	30	69	1,200,000
	American Magazine	30	61	1,000,000

Source: Mobil Oil, "Evolution of Mobil's Public Affairs Programs 1970–81" (Fairfax, VA, 1982)

Table 4.2. Mobil's Public Service Announcements by Year and Theme, 1975–1981

Year Aired	Theme
1975	Use mass transit
	Save gas – use car sparingly
	Car pooling
	Tune up car for better gas mileage
	Obey 55 mph speed limit
	Home-heating conservation
1976	Mobil Bicentennial posters – support your local museum
	Car care for summer driving
	Tire safety
	Winter tune-up
	Free enterprise system
1977	Offshore drilling
	Freedom of speech (including corporate)
	Law of supply and demand
	Big is not bad
1978	Offshore drilling
	Write your congressman
	Environmental protection
	Importance of deregulation
1979	Importance of industrial research (less regulation and taxes needed)
	Importance of national energy policy
	Importance of economic growth
	Welcome to spring – car care
1980	Election year – work for the candidate of your choice
	Importance of profits – in conjunction with Junior Achievement
	Importance of industrial research (less regulation and taxes needed)
	Importance of free market system
1981	Importance of economic growth
	Importance of domestic oil production

Source: Mobil Oil, "Evolution of Mobil's Public Affairs Programs 1970–81" (Fairfax, VA, 1982).

company was not allowed to advertise on the channel, it created a tagline that was voiced over the show each time it aired: "Made possible by a grant from Mobil Corporation, which invites you to join with them in supporting public television."[76] As media critic Laurence Jarvik notes, Mobil's PBS affiliation earned it considerable legitimacy in other realms where public relations was required. It greased the wheels of their lobbying efforts against oil company divestiture, and helped them counter President Jimmy Carter's National Energy Plan of 1979.[77]

Such efforts, with Mobil at the helm, "reflect a concerted effort to symbolically establish the corporation as a viable citizen in modern democracy."[78] Public relations had dramatic effects on the ways that corporations retooled the notion of public interest in their image. More to the point, this influence and these efforts shaped the way we understand environmentalism today. The messages that were communicated, such as those that balanced energy needs and economic growth, conspicuously avoided any mention of the environmental hazards of their actions. Indeed, the environment was painted as secondary to energy in this corporate-political discourse. The reasonable path, as we saw in chapter 3, was to focus on energy and economy. Activists pushing for policies and regulation that put environmental needs first were increasingly painted as unreasonable, irrational, and extreme.

Business in the Public Interest

Two more rule-of-reason-based strategies by public relations counselors would be of consequence for the transfer of business interests into the public interest. One was the portrayal of corporations as activists in and of themselves. This strategy was devised to express devotion to the spirit of public advocacy while in practice toppling the pedestal on which environmentalists had been perched. The other was the proliferation and institutionalization of industry–environmentalist partnerships to further entrench a consensus logic into environmental problem-solving.

Now that PR agents had seeded opportunities for clients to present their environmental commitments to concerned publics, they encouraged them to develop a more sustained program of communication, to anticipate environmental problems and become leaders in solving them. Like many of the other initiatives proposed by PR agents in this era, this was a means to promote their own competence as much as to burnish their clients' image. PR counselor Howard W. Chase was one of the more vocal proponents of this idea, which he termed "issue management." Issue management envisioned a systematic approach to information, one that not only communicates preestablished ideas but also forms them; that not merely manages environmental objectives but also anticipates and constructs them. Rather than asserting that the values of the corporation are in the public interest, he argued, the PR professional ought to *create* the public interest by helping to direct and indeed make public policy.[79]

Increasingly, articles about corporate political involvement characterized companies as "activists" in their own right.[80] Writing in the *California Management Review,* business professor S. Prakash Sethi described an evolutionary process by which companies became "activist" organizations to influence public policy. Companies should move from (1) a defensive, adversarial mode devoted to maintaining the status quo, past (2) an accommodative mode engaged in short-term campaigns in response to external factors, into (3) a stage of "positive activism." The positive activism mode involved long-term strategic planning "on the basis of a normative concept of 'public interest' and 'policy agenda' supported by the corporation." In this mode, senior management moved from "informal and secretive lobbying of key legislators" to "speaking out on public issues and offering advice and assistance to executive and legislative branches [of Congress]"; from non-controversial community affairs and corporate contributions to the "development of new groups . . . in support of a national policy agenda"; and from resistance to other groups' viewpoints to "emphasis on the development of third sector as bulwark against increasing government encroachment in the social arena" as well as public communications and education to advocate specific policies and programs.[81]

"The essence of corporate political activism," Sethi concluded, "is for the corporation to develop a cogent view of the public interest and, then, political positions and strategies that embody this notion."[82] Corporate communicators helped their clients become "activists" by adopting not only the title but also the techniques of public interest groups, such as coalition building for indirect (grassroots) lobbying. This approach caused the director of one of Ralph Nader's research groups to complain to the *National Journal,* "[Business coalitions] have taken the techniques, such as working with the press and grass roots, that we've been successful with, but they do it better because they have more money and manpower."[83]

One way business leaders tried to get out in front of the environmental issue was to create forums for dialogue, in the spirit of "cooperative pluralism."[84] Could "producer groups," such as coal companies and electric utilities, interact with "countervailing power groups," such as environmental advocates, without government involvement in order to negotiate and seek consensus around matters of public policy? Some saw a productive answer in the National Coal Policy Project (NCPP).

The immediate background of the NCPP was the desire by industry to influence American domestic energy policy in the aftermath of the 1973–1974 oil crisis. But a broader postwar context is more instructive. Coal-fired power plants had approximately doubled their sulfur oxides emissions every decade between 1940 and 1970.[85] Throughout the 1960s, coal producers (and consumers) as well as electric utilities opposed any government regulation of air pollution. This "coal coalition," as historian Richard Vietor describes it, prevented amendments to the Clean Air Act in 1963 and stymied federal emission standards for industrial air pollutants in 1967. It was not until 1970 that the new Clean Air Act finally gave the federal government the authority to control air pollution.[86]

The NCPP was industry's effort to retrieve some control over policy-making and over air pollution standards. Its stated purpose was "to bring together individuals from industry and environmental organizations for the purpose of achieving a consensus on a detailed plan to permit the responsible use and conservation of coal in an economic and environmentally acceptable manner." Over the course of the five-year project, a series of meetings were held between environmental action groups, coal mining executives, and industrialists to find areas of compromise through the exercise of reason. Journalists were invited to observe, as were (on a limited basis) government officials. The participants were enjoined to "avoid . . . the lawyers' standard tactics based on deceit, ad hominem attack, procedural devices and delays— tactics designed to win by any means—tactics that do not serve the public interest."[87]

The NCPP's 1978 report, tellingly titled, "Where We Agree," dedicated over 800 pages to "narrow[ing] the policy differences separating environmentalists from the producers and consumers of coal." The ultimate outcome of the NCPP was not, however, to transform policy on coal but to give PR executives another form of justification and publicity for their objectives. As business professor Reed Moyer wrote in a review of the report that same year, "This work's greatest value . . . is not necessarily its informational content. Rather, it is perhaps most important for its delineation and sharpening of issues separating environmentalists and industry representatives and for its creation of a model for conflict resolution, the adoption of which could profit other adversarial groups."[88]

PR people were paying close attention. An editorial in *Chemical Week* cited the NCPP as an overture to "a 1980s era of cooperation."[89] *Chemical & Engineering News* noted that while the project had not had the anticipated impact on federal policy decisions around coal, it had nevertheless shown the value of "reason and mutual respect to find areas of agreement" which may influence decisions down the road.[90] E. Bruce Harrison, writing in *PR Journal*, said the NCPP "puts a fresh light on fair play as a way to solve legal and public relations problems of the corporation."[91]

The 1980s era of "cooperation" was indeed at hand, notably in the proliferation of industry–environmentalist partnerships. One reason, paradoxically, for the success of the partnership model was that it seemed more oriented toward accommodation of different viewpoints than did the earlier tactics of environmental advocates. Increasingly, the path of litigation was obstructed by industry's "voluntary" efforts to deal with pollution problems in a transparent way, and the accompanying publicity effects of its efforts. Initiatives by environmentalists to protest economic growth at the expense of environmental protection were painted as anti-progress, backward-looking, and unrealistic. Second, as environmental organizations set up offices in Washington, a different kind of compromise took place. Grassroots activists were no longer able to work at the grassroots, having to play by the "rules of the game" in Washington.[92] Some of the more conservative or "apolitical" environmental organizations, such as the National Wildlife Federation, established partnerships in an effort to balance environmental goals and economic growth. The Federation's Corporate Conservation Council, established in 1982 and made up of executives from seventeen major corporations, aimed to transfer managerial and technical skills to public sector actors. "Public sector managers who combine the 'stick' of traditional 'command and control' regulation with the 'carrot' of profitable business opportunities offered by environmental protection will be better able to carry out their jobs."[93]

The case of the Nature Conservancy reveals the degree to which business had gained legitimacy in the environmental realm. The Nature Conservancy was at that time the only nonprofit national conservation organization devoted exclusively to land preservation. Its methods were rooted in trying to persuade owners of ecologically important land parcels to either sell their land to the Conservancy for subsequent sale to the government or by arranging direct transfer. According to the Conservancy, in 1978, over 100 million acres of land in the United States were owned by twenty major US corporations.[94] The conservancy began to realize in the 1970s that appealing to the company's economic (tax-deductible) incentive of transferring their land could be amplified by the reputational dimensions of acting in the public good.[95]

The organization created an extensive public relations campaign, including newsletters and brochures, press releases and business-media relations, slide presentations and a short film, and special ceremonies to honor corporate land contributors. Painting itself as a pragmatic, compromise-seeking, and industry-allied environmental group, the Conservancy gained the favor of company leaders. As one magazine article explained,

> This is a different breed of environmentalist. The Conservancy doesn't speak of a corporation's questionable environmental planning or of its sins against nature. It speaks instead of . . . the reasons why conservation makes good economic sense . . . This new approach, free of emotional pleas and threats of legal challenges, has paid off. The Nature Conservancy now claims 105 American corporations as paying members in the organization, and it boasted assets of $100 million in 1976.[96]

One executive said, "past experience in working with the Conservancy to develop realistic conservation projects throughout the nation has assured us that this project would be completed in the public interest."[97] Wary of being characterized as a "sell out," the Conservancy instead highlighted its relationship with business as a positive opportunity in which openness and dialogue could lead to benefits for all concerned. In 1979, the Nature Conservancy presented its case study to the Public Relations Society of America, winning the association's "Silver Anvil" award for effective PR (figure 4.2).[98]

"We can
work with you..."

Figure 4.2. "We Can Work with You." Nature Conservancy, 1978. *Article source:* Peter Wood, "Business-suited Saviors of Nation's Vanishing Wilds." *Smithsonian*, December 1978. Photograph by Yoichi R. Okamoto. Reproduced with permission by the Okamoto family.

Conclusion: Compromising the Environment

When, in 1990, President George Bush Sr. signed into law new amendments to the Clean Air Act, PR counselors saw it as a hard-won victory: "The result," as E. Bruce Harrison put it, "of more than a decade of compromise between government, environmentalists and industry." By 1991, the federal government would establish the President's Commission on Environmental Quality (PCEQ), cementing the legitimacy of public–private environmental partnerships. "The idea behind PCEQ was to find a way to replicate on a wider basis the success of certain private sector initiatives in economically protecting the environment, explained Michael Deland, chairman of the Council on Environmental Quality in the Executive Office of the president. "We asked people with proven successes in industry to join with members of the environmental, foundation, and academic community, to get them working in common cause. These were people who if they had communicated before, it probably would have been through lawyers in a courtroom."[99]

In "An Obit for an -Ism," an opinion piece in the public affairs newsletter *Impact* in 1992, Harrison argued that industry had "become the managing partner of the environment," with business "taking possession of greening." For Harrison, this moment signified "the death of environmentalism," a political concept no longer needed in the last decade of the twentieth century. History would show Harrison to be wrong. But he was right about one thing: the decade of compromise, consensus-making, and collaboration would prove devastating to the promotion of the environment as a public problem.

5

Sustainable Communication

Green PR and the Export of Corporate Environmentalism

By most accounts, the United Nations Conference on Environment and Development (UNCED) in Rio de Janeiro in June 1992 was a failure. The symbolic potential of the event—delegations from 178 countries, heads of state of over 100 countries, and more than 1,000 nongovernmental organizations assembled to reinforce common cause and international laws around environmental protection—was not realized in practice. Despite civil society calls to embrace the principles of the environmental movement, mounting evidence of anthropogenic causes of climate change, and clear indications of the outsized role of industrial activity in perpetuating environmental disasters, debates at the Earth Summit—as the UNCED event was also known—were not transformed into enforceable regulation.

It may have had something to do with the expectations that surrounded the conference. Twenty years earlier, the 1972 United Nations Conference on the Human Environment (UNCHE) in Stockholm had been catalyzed by increasing attention to the causes of environmental degradation worldwide. The significance of the conference was manifested in the explicit call for, and subsequent articulation of, a common approach to reckoning with planetary ecological systems in a sustained and coordinated way. The UNCHE led not only to greater global awareness and responsibility-taking but also to the formulation of international environmental norms and laws, as well as institutions dedicated to enforcing and monitoring these new standards. In particular, the United Nations Environment Programme (UNEP) was created as a permanent agency, a global body that would act as "the environmental conscience" of the United Nations system and coordinate with other key agencies such as the World Health Organization.[1]

By the time the UNCED rolled around in the early 1990s, attempts to enforce a coherent global action plan around the environment had exposed fault lines among different parties to the agreements. The rift was particularly

A Strategic Nature. Melissa Aronczyk and Maria I. Espinoza, Oxford University Press. © Oxford University Press 2022.
DOI: 10.1093/oso/9780190055349.003.0006

wide between North and South, as the Brundtland Report made abundantly clear.[2] Developing countries felt they had to bear an unfair burden relative to their developed world counterparts and resented the yoking together of environment and development issues without concrete regard for pressing concerns such as unemployment and urbanization.[3]

Business leaders were also disgruntled. As we have seen in earlier chapters, American companies had been considerably decentered in the late 1960s and early 1970s. The emergence and consolidation of an "age of activism," which married full-throated calls for action around environmental issues (among other social and political concerns) with new legal supports and regulatory institutions, had created a "PR crisis" for business, leading companies to retrench in "survival" mode. The 1972 Stockholm Declaration on the Human Environment that followed the UNCHE articulated norms and principles that would form the basis for (soft) international law. These principles enjoined countries to adhere to conservation and redistribution of renewable and nonrenewable resources and emphasized countries' responsibility for environmental damage caused beyond their jurisdiction. Crucially, the Declaration privileged government policies to enforce environmental protection over market solutions.[4] To the extent that industry was included at all in the UNCHE negotiations, it was as a culprit and a threat.

The events leading up to and after the 1992 conference tell a dramatically different story. In the intervening twenty years between the two UN conferences, business developed extensive means of interacting with the concept and practice of environmentalism. Rather than reenforcing a global consensus on norms of environmental protection, the UNCED revealed the extent of industrial integration into environmental institutions, norms, and practice. If this integration was in part an outcome of concerted organizational transformation—within both private and public sector institutions—it was also the result of a clear conceptual shift in the meaning of environmentalism, hinging on the notion of sustainability. Indeed, as several scholars have argued, it was the rearticulation—what Leslie Sklair calls the "capture"—of sustainability by corporate actors that led to the consolidation of the business community as a fixture in both the policymaking and the pragmatic response to the problem of global environmental destruction.[5]

The reframing and institutionalization of "sustainable development" as a mainstream response to environmental concern was a paradigm shift. It reanchored environmentalism as a feature of a liberal international economic order, an order that celebrates growth as its central tenet and sees market and economic mechanisms as endemic to this growth.[6] This model embraces voluntary, rather than government-imposed, norms; and systems of regulation to "manage," rather than "control," environmental change.[7]

It is relevant that Bernstein refers to this transformation as the "compromise" of liberal environmentalism. The process of "compromising for the common good," as Boltanski and Thévenot explain, is a dialectics of continual justification and critique. As we saw in chapter 4, when parties to a debate or contestation advance different visions for organizing social life, they search for a higher ground on which agreement can be reached, achieving an outcome that "sounds right" to all concerned. Paradoxically, it was just such a compromise, contained in the "artful vagueness" of the term "sustainability," that was the source of its widespread support.[8] "Sustainability" offered the unified theme that had been lacking at the UNCHE. In its openness to interpretation by multiple parties—countries in the North and South, social movements and bureaucrats, international and local organizations, and capitalist and socialist regimes—the idea of sustainability allowed a higher principle of agreement to be reached. Sustainability created a spirit of international cooperation in environmental protection. And it was this spirit of cooperation, enabled by a series of interpretive and relativized techniques and expertise, that would be embedded in the next few decades of international responses to and action around global environmental problems.[9]

The question that remains is how this spirit of compromise was embedded into institutional practice. In 1992, the meeting at Rio seemed to crystallize a vision of environmentalism that had been accelerating throughout the 1980s, in which managerial, technical, and financial resources are appended to the resolution of environmental challenges. Although several studies have made clear *why* the set of ideas associated with liberal environmentalism has become institutionalized, they have told only part of the story of *how* this process took place in the events leading up to and following the conference at Rio. This chapter answers the "how" question by paying close attention to the efforts of public relations. It shows how PR actors played instrumental roles in creating the spirit of compromise that supported the creation of sustainable development programs in international settings. We track the formation and evolution of a specialized field of "green" public relations in which particular ideas about the environment were conceptualized, stabilized, and circulated as tools of international environmental governance. Environmental PR was deeply invested in creating the cultural categories that rooted the ideas, power differentials, and patterns of interaction of sustainable development, and in shaping the information environment in which these categories and practices could flourish.

In multiple forums and dialogues leading up to the UN conference in Rio and at the Rio conference itself, public relations actors played crucial co-ordinating roles. After Rio, the values and meanings of sustainability were further embedded, institutionalized, and circulated internationally through the efforts of public relations networks. We examine the emergence of this network and the organizational forms and practices in which it becomes enmeshed as well as the content of its claims, showing how American PR action around international environmental governance created the ideological conditions for diffusing "green communication" and thus for championing a particular "brand" of environmentalism overseas, one rooted in voluntary, strategic, and entrepreneurial approaches to environmental problems.[10]

To get a handle on what happened during the twenty-year interval between the two United Nations conferences to allow public relations to gain a foothold on the terrain of international environmental governance, we consider a book that had a dramatic impact on public perceptions of environmental knowledge: *The Limits to Growth*. Though this book is not in the least about public relations, it offers a perspective on how certain ways of knowing the environment came into being in the latter third of the twentieth century, and how these ways of knowing impacted the climate of publicity around environmental problems, which would in turn be taken up by public relations experts.

Limits to Growth and a Prehistory of Corporate Environmentalism

A defining mantra of the UNCHE in 1972 was that the environment was a finite resource. In emphasizing the relationship between control of the environment and patterns of development, the conference participants (114 government delegations) advanced the understanding that current patterns of economic, demographic, and industrial growth were not sustainable. For many of the participants, especially those in developed countries, the takeaway was that countries needed to make a choice between economic and environmental growth.[11]

This takeaway was expressed in no uncertain terms in a report published the same year called *The Limits to Growth*. Still today the bestselling environmental title ever published, with 12 million copies sold, the report heralded a permanent shift in the grounds of environmental awareness and action in both public and political spheres.[12] *Limits to Growth* was published to great fanfare in early March 1972, preceded (ironically) by a dedicated public relations campaign orchestrated by Potomac Associates, a for-profit publishing venture run by friends of the report's authors. As Hecox describes it, the PR campaign was designed to generate interest in the book's dramatic findings and to ensure that the book was "placed in the 'right' hands so that its message could influence policy and stir public debate."[13] Through the promotion, networking, and media attention to the book's argument and model, not to mention its translation into thirty-seven languages, the book made its mark on the climate of publicity.

Limits to Growth centered environmental concern and policymaking around the depletion of nonrenewable resources, the human-centered causes of environmental degradation, and the global character of response.[14] It revealed the fundamental interdependency of environmental, economic, and energy systems and the impact of overusing those systems. Employing a system dynamics approach, the report used computer modeling to project a future scenario of total environmental deterioration if current patterns of industrial growth, energy consumption, and world population were not abated. In its dire predictions of "overshoot and collapse," the report both legitimated the use of technical data as a means of accurate forecasting of social outcomes and entrenched a sense of urgency among policymakers and citizens.[15]

The book also represented a paradigm shift in the character of environmental knowledge and its knowers. The report was produced by the Club of Rome, an international group of business leaders, politicians, and scientists organized by the Italian businessman Aurelio Peccei. The simulation model that aggregated global data to inform the report's central argument was called a "world dynamics" model, invented by Jay Forrester, professor and computer expert at the Sloan School of Management, MIT (Massachusetts Institute of Technology). In 1970, the Club of Rome had been struggling to develop a methodology that would allow them to address what they called a "world *problématique*"—large-scale, persistent, complex problems common to all societies—by identifying the interrelated technical, social, economic, and political factors that cause them. Carroll Wilson, a colleague of Forrester at MIT and a member of the Club of Rome, brought Forrester in to help them.

With the Club of Rome, Forrester assembled a team of researchers known as the System Dynamics Group to apply his world dynamics model to a complex of five interrelated macro variables in the *problématique*: "accelerating industrialization, rapid population growth, widespread malnutrition, depletion of nonrenewable resources, and a deteriorating environment."[16] Forrester's belief was that all systemic problems followed a common progression: there are stocks, there are flows, and there is feedback. Stocks are quantities of material or information that accumulate over time. These stocks are affected by flows—the movement of this material or information into and out of the system. The feedback, also a source of information, regulates the stocks and flows to achieve states of balance. Systems have limits; exceeding the limits is known as overshoot; and overshoot can lead to collapse.[17]

Forrester's model had seen some prior testing and application on engineering and management problems. In the 1950s and 1960s, Forrester had applied systems modeling on radar and combat information for the US military. At MIT, his role was to apply lessons learned to organizational problems in corporate management. For the project envisioned by the Club of Rome, the problem was far more complex. Here was a global problem of interdependent ecological, demographic, socioeconomic, and political systems, with parameters that far exceeded the boundaries of managerial or engineering thinking. More critically, as Paul Edwards has shown, it was more or less impossible, in the early 1970s, to accumulate the kind of global, longitudinal data required to accurately forecast a planetary environmental future. Potential sources of cross-national information, such as the United Nations or the World Bank, were limited by their lack of long-term findings, as these organizations were barely twenty-five years old at the time. In any case, Forrester was far more interested in the model than he was in the data. It was the structure and dynamics that mattered to Forrester, not the specific inputs. The data collected to make the argument in *The Limits to Growth* were therefore incomplete, inaccurate, and wildly incommensurate across space and time. Most scientists refused to accept the premises of the world dynamics model, and considerable questions emerged about the report's conclusions. Perhaps the most damning critique of the report, emerging shortly after its publication, was that "the modeling techniques themselves—not the phenomena supposedly modeled—generated the characteristic behavior."[18] In other words, no matter what data was entered into the system, the system would always show the same pattern of stocks, flows, and feedback, with a strong potential for overshoot and collapse.

Seen from a distance of fifty years, what remains most impressive about *The Limits of Growth* was that massive flaws in the data didn't really matter in the long run. Rather than the quality of its data or the accuracy of its predictions, it was the compelling power of its narrative that explains the book's longevity as a cultural and political resource and its tremendous ability to influence the character of environmental knowledge. Taken as a heuristic—or what we have been calling in this book a technology of legitimacy—*The Limits to Growth* was an unqualified success.

There are at least three ways in which *The Limits to Growth* authenticated and laid the groundwork for the information and influence methods that would allow PR actors to promote sustainable development worldwide leading up to and after Rio.

First was the idea that it was now both necessary and possible to assemble truly *global* environmental data. One of the desired outcomes of the UNCHE was "an international consensus on an environmental ethic and on the basic principles that should guide the environmental relationships of the international community."[19] Like other so-called global environmental information systems established at this time, such as the UNEP's Infoterra database, the world dynamics model of *The Limits to Growth* offered the potential for this international consensus.[20] A global system of information that relied on technical, systemic models seemed to redirect environmentalism away from the political and economic concerns that reflected specific national priorities.

Second, *The Limits to Growth* helped justify a *future* orientation for environmental problems. Economic futures research and forecasting had gained popularity throughout the 1950s and 1960s. In the 1970s, futures research evolved in two different and seemingly opposite directions. The first was a result of advanced computing, which enabled the kind of simulation modeling Forrester and his team deployed for the book. Such modeling practices not only helped establish predictive technologies as legitimate tools for policymaking, but also contributed to establishing legitimacy for global warming itself as a policy issue.[21] Modeling helped turn a seemingly invisible, abstract phenomenon into something detectable, predictable, and knowable.

The other path for futures research took place within organizations. Companies and government offices undertook futures research to gain insight into patterns of social and political life, developing methods in public opinion measurement, long-range and scenario planning, and issue management to anticipate and control problems affecting their operations.[22] This would come to matter a great deal for companies seeking strategies to offset the impact of environmental concerns before they became stabilized as policy. So even as thinking about the future helped to concretize environmental issues such as global warming, it also allowed actors to develop repertoires of contention to push them off.

Third, *The Limits to Growth* opened the door to the acceptance of industrial knowledge as a corollary to scientific expertise around environmental issues: Aurelio Peccei, director of the Club of Rome, was a prominent industrialist, and Jay Forrester, in his early days at MIT, had applied systems thinking to management problems for companies like General Electric— credentials that enhanced this acceptance. But perhaps the book's greatest impact on the legitimacy of modern industrial knowledge around environmentalism lies in how business responded to its central thesis that current patterns of existence were unsustainable. The kind of sustainability that *Limits* espoused—balanced, equitable, restrained, conservationist—was anathema to the principles of industrial and economic growth.[23] In the decades after the book's publication, the business community would turn considerable resources toward anticipating and controlling this new opponent, developing new models and systems to integrate industrial priorities with environmental ones.

To fully comprehend the ways that sustainability became a watchword for firms, we need to understand the work of PR. Using the environmental inputs made manifest in *Limits*—technical data at a global scale, systems and futures thinking, and international consensus—public relations experts would work to embed their own knowledge into the capacious concept of sustainability.

EnviroComm and Epistemic Authority

A key theme we have been examining throughout this book is what it means to consider public relations experts as an epistemic community. Studies on environmental governance have frequently employed the notion of epistemic communities "since the complex and uncertain nature of environmental problems appears to privilege experts in determining the nature of environmental problems and the technical requirements needed to address them."[24] Defined as self-structured groups sharing professional expertise, beliefs, and common objectives for influencing public policy, epistemic communities claim authority over expert knowledge and seek to embed this legitimacy

into their objectives.[25] Despite their lack of de jure authority, epistemic communities can shape future policy development by defining the issue at stake
and providing standards or normative guidance that is not otherwise available. As "knowledge-based networks," epistemic communities also influence
meaning-making processes by circulating particular understandings of issues among different publics.[26]

Although the notion of epistemic communities was developed around
the idea of scientific expertise, the issues surrounding global environmental
governance require an expanded idea of what constitutes "knowledge" and
the status of the "knowers." As David Levy and Peter Newell describe it, the
term "environmental governance" signifies "the broad range of political, economic and social structures and processes that shape and constrain actors'
behavior towards the environment."[27] In this understanding, environmental
governance is not limited to rule-making and enforcement or the creation of
institutions but encompasses "a soft infrastructure of norms, expectations,
and social understandings of acceptable behavior towards the environment,
in processes that engage the participation of a broad range of stakeholders."[28]
In the highly contested arena of environmental politics and its publics,
other kinds of experts, such as business networks, think tanks, international
foundations, and multinational consultants gain entrée into the debate.
Diane Stone argues that these groups facilitate the exchange of knowledge
among decision-makers across borders, using "their intellectual authority or
market expertise to reinforce and legitimate certain forms of policy or normative standards as best practice."[29]

Of course, this knowledge is never neutral. Private sector networks tend
to promote "the globalization of the core values of Western culture generally, and the transmission of the idea of liberalization specifically," acting
as "reputational intermediaries" that signal to a wider international audience of investors and financial institutions that a country is a safe, "normal"
place in which to do business, that the "right" people are involved, and that
decisions made at the local, regional, or national level will resonate with
global expectations.[30]

Such is the case for EnviroComm, a network of public relations and public
affairs firms created in the late 1980s in Washington, DC, that grew over the
next ten years to include PR firms across Europe as well as in Mexico. While
it was not the first international network of public relations companies,
EnviroComm was the first network to focus exclusively on "green" PR and on
disseminating environmental expertise among its members and clients. In

its ability to create standardized, predictive, and industrially savvy information about the environment and to circulate this in international networks, EnviroComm represents a textbook case of the kind of environmental knowledge made possible by *Limits to Growth*—even as this knowledge was mobilized deliberately to counter the book's conclusions. The network helped intensify green communication across borders and among contentious industries (including tobacco, fossil fuels, and chemicals), transforming green PR from a specialized skill into part of the "dogma" of environmental management.[31] As we shall see, EnviroComm's approach was focused less on the policies themselves than on the means by which certain forms of governance can be made to appear more legitimate than others.

Making the Corporate Environment: EnviroComm and the Creation of Sustainable Communication

In 1989, the environmental public relations firm E. Bruce Harrison Co. entered into a partnership with the Brussels-based public affairs consultancy, Andersson Elffers Felix (AEF). The choice of Brussels was strategic. The regional integration of "the world's biggest marketplace" in Europe, along with new European Community standards for environmental protection, suggested that, as the page A1 headline in the 17 May 1989 *Wall Street Journal* put it, "European bureaucrats are writing the rules Americans will live by."[32] For American corporations, the consolidation of the European Single Market held both promise and potential peril. In terms of promise, US firms could help European companies learn the ropes of environmental compliance. American companies had already been exposed to environmental controls by federal bodies such as the Environmental Protection Agency in the context of right-to-know legislation and the Clean Air Act debates. In terms of peril, many of the European proposals for environmental standards, such as eco-labeling and emission restrictions, were much more demanding than their American counterparts. American companies with units abroad sought advice on how to navigate these new rules along with their regulators.

AEF/Harrison International (the partnership would become known as EnviroComm in 1994), was designed to provide "an early warning system" for client companies, monitoring as of 1989 up to eighteen "agent institutions" involved in making the European Community's environmental policy and sending regular reports back to Washington.[33] In addition to monitoring environmental policy for clients, AEF's role was "to influence [policy] creation" through strategic lobbying. Based on the list of institutions in the AEF report, we may surmise that the monitoring was extensive. Trade groups and workers' unions, municipal and state offices, regional chambers of commerce, and departments in the European Community's legislative and executive branches were all under scrutiny[34]:

> European Commission (EC)
> Directorates General
> EC representatives in the member states
> Council of Ministers
> European Parliament
> Economic and Social Committee
> European action and pressure groups
> European employers' and workers' organizations
> Permanent representatives at the EC
> EC liaison offices
> EC advisory centers
> National departments and governments
> National political parties and people's representatives
> Regional chambers of commerce
> Regional and municipal administrations
> Diplomatic missions in the member states
> Employers' and workers' organizations in the member states
> EFTA (European Free Trade Association) countries[35]

As for any other expert network, EnviroComm's capacity to make authoritative claims over environmental communication rested in part on its members' reputation and experience in the field. In this case, the establishment of environmental communication as a field of its own was directly linked to the reputation of EnviroComm's founder: the American firm E. Bruce Harrison Co. Harrison had been working for decades to enforce the principles of "green" PR. By the late 1980s, the firm was highly successful, consistently ranked by O'Dwyer's (the leading trade publication) among the top ten environmental public relations firms in terms of billings. In 1992 the firm claimed to represent "through coalitions and direct service . . . more than eighty of the Fortune 500."[36]

With the understanding that shaping environmental policy required not only political but also cultural influence, AEF/Harrison International organized and participated in several public events to present its experience in managing environmental affairs for corporate clients. While positioning itself as a source of expert knowledge in the European context, AEF/Harrison International urged companies to integrate a green PR component into their environmental management activities.

To demonstrate the value of its expertise abroad, AEF/Harrison International created and sponsored media events designed to raise the visibility and legitimacy of the firm's offerings. One such event was the international promotion of a newly published book by a Harrison Company vice-president, Ernest Wittenberg, and his wife Elisabeth Wittenberg, *How to Win in Washington: Very Practical Advice about Lobbying the Grassroots and the Media*. While domestic promotion focused on building up the Wittenbergs as super-connected Washington insiders,[37] the European coverage was framed to highlight the growing similarities between American-style and European public affairs. The Brussels-based financial magazine, *Trends*, called *How to Win in Washington* "a blockbuster at the Berlaymont" and claimed, "In one evening [of reading the book] you will learn how to get the eye and the ear of lawmakers in Brussels, Luxembourg and Strasbourg."[38] Wittenberg himself also wrote op-eds in the *New York Times* and elsewhere underlining the growing industry for US public affairs in Europe.[39]

A second means of promoting green PR in Europe was to relay the message that companies that do not seek representation for environmental issues in Europe ran considerable economic and political risk. One article sums up the general mood:

> French producers of mineral water forward the idea that Bonn's decision to offer a rebate for plastic PVC bottles as a form of environmental protection is in fact a form of disguised protectionism. Tobacco producers fight against a ban on TV commercials for cigarettes. . . . [P]rinters worry that new antipollution laws will make certain solvents unusable. . . . They are all hurrying to Berlaymont, the headquarters of the European Commission, with the same obsession: To advance their cause (law, regulation, financing request, complaint). Their credo: "What happens without us risks working against us."[40]

As a third means of building epistemic authority, Harrison sought membership in and stewardship of international organizations. Indeed, throughout the 1980s, Harrison had been setting the international stage to perform his expertise. In November 1984, he attended the first meeting of the World Industry Conference on Environmental Management (WICEM I) at the Palais des Congrès in Versailles, France. Organized and hosted by UNEP in cooperation with the International Chamber of Commerce (ICC), the conference was meant to "move relations between world industry and environmental organizations from confrontation to cooperation." In the intervening years since the 1972 UNCHE conference, UNEP's role had evolved from acting as the environmental conscience of the UN to playing peacemaker among dissatisfied stakeholders in the environmental policy sphere. WICEM I was in effect an olive branch from the UN to industry in an effort to encourage greater participation by companies in international and intergovernmental environmental policy. As the executive director of UNEP, Mostafa Tolba, told the assembled WICEM I delegates in his opening address, "It is not too much to hope that successful cooperation on environmental matters could lead to an improvement in the climate for negotiations in other areas of the global *problématique*."[41] Unlike the "limits to growth" paradigm set in the 1970s, sustainable growth was presented at WICEM I as a principle with which industry could find accommodation. Harrison, along with the industrial delegates to the conference, took careful note.[42]

When, a couple of years later, the ICC formed an International Environmental Bureau, its founding members had all attended WICEM I.[43] The bureau saw as its mission to take control of the sustainability paradigm and turn it to industry's advantage. The strategy at root, in which Harrison would participate, was to create a network of like-minded organizations and initiatives that would duplicate and reinforce the efforts of the ICC. Working in tandem, these groups would develop standards and protocols to account for and promote industry's environmental activities. By creating their own benchmarks, audits, codes of conduct, and certification programs, industrial actors developed an alternative framework of knowledge through which to publicize their actions as "sustainable."[44] The success of corporate environmentalism as a technology of legitimacy for the ongoing activities of polluting firms is in part, as we have seen, a matter of the connotative indeterminacy of the notion of sustainability and its availability for capture by different actors with varying motivations. Key to the success of this strategy, however, was that the framework of knowledge these actors created was built according to the same parameters developed in *Limits to Growth*. If *Limits* had articulated environmental knowledge as global, future-oriented, and information-based, the industrial response was to adopt the same conditions of knowledge and mold them to their own more or less diametrically opposed ideological purposes.

The promotion of corporate environmentalism relied therefore on disciplined, rule-bound, and highly managed communication strategies to maintain a common language and unity of purpose about the benefits of economic growth as aligned with the needs of environmental protection. And this was the arena in which E. Bruce Harrison was most qualified. Over the next decade, EnviroComm would operate as the hub of a corporate environmental network that included the ICC and its "sister" organization, the United States Council for International Business (USCIB); the Global Environmental Management Initiative (GEMI), a coalition of around twenty companies formed and overseen by E. Bruce Harrison's firm; the Chemical Manufacturers' Association and its self-certification and environmental compliance program Responsible Care; and other satellite groups such as the World Environment Center.

The target event for the launch and publicity surrounding this corporate environmentalism was the UNCED conference in Rio. This international UN conference was not just a means of capturing international attention and approbation for this project; it also was meant to show the approval of governments and intergovernmental organizations for this version of environmentalism. The second WICEM gathering, in Rotterdam, was planned to take place just ahead of the UNCED conference and was co-sponsored by UNCED and UNEP. In December 1990, the WICEM II preparatory working group met in Zurich for a "brainstorming session." Co-chairs of the working group were Union Carbide CEO Robert Kennedy and president and CEO of Dow Chemical, Frank Popoff, both clients of Harrison. Harrison was a member of the working group, as was Norine Kennedy of the US Council for International Business. The notes from the brainstorming session make the group's objectives clear:

What are the main messages? What do we want to achieve?
On enterprise level: 1. Regain credibility: Industry has achieved much and is not the main source of the env. problem any more. We can and want to prove it. Leading corporations are actually willing to go quite much further in their own operations than what the public believed, and are taking joint action to do so. 2. Take the lead in the most critical issues: Industry has the competence to solve the major environmental challenges and is taking

the initiative. **On political level:** 1. To work within the framework of the MARKET ECONOMY which requires: carefully planned use of economic instruments; reasonably "level playing field" that does not distort int. trade. 2. To PARTICIPATE and have an influence on development of national-regional-global policies which requires the public/governments to accept that Industry and Commerce are the key to SOLVING the environmental and development challenges."[45]

The solution the working group envisioned took the form of a Business Charter on Sustainable Development. The charter would stabilize and help circulate the main message at WICEM II. As the group decided, "World business launches a specific initiative to induce an attitude change and a new 'sustainable development dimension' in corporate life: THE CHARTER" and "Business will go through a change of resource perception, and will also gain credibility by actively implementing the charter." The working group also developed plans for international energy cooperation schemes to address global warming; and new models for technology transfer (financial and information technologies). The Business Charter was intended to act as a rival object to more formal and binding agreements. The goal was to develop a version of environmentalism into which companies could insert themselves without being subjected to the restrictive regulatory frameworks advanced at the UNCHE and embedded in environmental legislation in the United States and abroad.

As the public relations representative for business leaders attending the summit, Harrison was centrally involved in the events leading up to the conference as well as the preparation of the Business Charter. Of the 203 companies and business organizations worldwide that had signed on to the Business Charter by March 1991, thirty-seven were American companies; and of these thirty-seven, more than half were clients of the E. Bruce Harrison Co. These were not minor players. They included some of the largest companies in the chemical sector: the Dow Chemical Company; E. I. Du Pont de Nemours & Co.; and the Union Carbide Corporation, among others.[46]

The Business Charter on Sustainable Development was given its first airing at WICEM II in Rotterdam. This event was as much a planning session for the public relations effort that would take place at UNCED as it was a rendering of international responses to the goals articulated at WICEM I. Since many of the speakers at WICEM II were Harrison clients, Harrison wrote or co-wrote many of the speeches delivered at the conference. Speeches by

Robert Kennedy at Union Carbide; David Buzzelli, VP and corporate director, Environment, Health and Safety at Dow; W. Ross Stevens III and E. S. Woolard at Du Pont; and Margaret Kerr, VP of Environment, Health and Safety at Northern Telecom and president of the Industry Cooperative for Ozone Layer Protection (ICOLP), all bore Harrison's stamp.

Three main ideas were advanced at WICEM II, ideas that were carried forward to the United Nations Earth Summit in Rio and encoded into the principles of EnviroComm. These were (1) to promote and prioritize voluntary, market-based approaches to environmental action by world industry; (2) to encourage the recognition of economic growth and environmental protection as mutually reinforcing endeavors; and (3) to advance the idea that business must play a key role in economic growth and in environmental protection.

The International Public Relations Association—of which Harrison was an active member and, as of fall 1993, chair of its Environmental Committee— also created a communications guide, the Nairobi Code for Communication on Environment and Development, which exhorted IPRA members to "accept that they have a responsibility to ensure that the information and counsel which they provide, and products or services which they promote, fall within the context of sustainable development."[47] The IPRA was careful to say, however, that "members shall seek to develop programmes which counsel and communicate on the benefit of a *balanced consideration* of environmental, economic and social development factors." Domestically, working group members and their networks gave speeches to the PRSA and internally within their own companies to ensure the message was disseminated.[48]

From the point of view of industry participants, the events at Rio were a success. Voluntary codes of conduct such as Responsible Care for the chemical industry, the Global Environmental Management Initiative for general environmental management, and the Business Charter were largely adopted by the international community as viable responses to the Brundtland Report. Chapter 30 of Agenda 21, the action plan devised after Rio, encouraged business and industry to "report annually on their environmental records," but such environmental audits were not formalized or legislated. The successful experience of US corporate leaders and their representatives at UNCED crystallized for Harrison the opportunity to promote such voluntary compliance programs and codes of conduct internationally, and the role of public relations in doing so.

It was also here that Harrison first developed the concept that would form the supporting pillars of his work with EnviroComm and after: sustainable communication. In a 1993 article for IPRA members, titled "Green

Table 5.1. Company Membership in "Sustainable Business" Networks, 1986–1999

Company	EBH Client[a]	CMA[b]	Responsible Care[c]	BCSD[d]	NEDA[e]	GEMI[f]	Charter Signatory[g]
Air Products & Chemicals, Inc.	X	X					X
Anheuser-Busch Companies	X				X	X	
Ashland Oil, Inc.	X	X	X (1997)		X	X	X
AT&T	X			X (1995)	X	X	X
BASF AG	X	X				X	X
British Petroleum (BP Chemicals)	X	X	X (n.d.)	X (1995)			X
Browning-Ferris Industries, Inc.				X (1992)		X	X
Chevron Corp.	X	X	X (1998)	X (1992)	X		X
Coca-Cola	X		X (1999)			X	
Colgate-Palmolive Co.	X					X	X
Coors Brewing (Adolph Coors)	X			X (1997)		X	X
Dow Chemical Co.	X	X	X (1999)	X (1992)	X	X	X
E. I. Du Pont de Nemours & Co.	X	X	X (1995)	X (1992)	X	X	X
Eastman Kodak Co. (Eastman Chemical Co.)	X	X	X (1998)	X (1997)	X	X	X
Elf Aquitaine/Elf Atochem North America		X	X (1997)			X	X
Exxon Chemical Co.	X		X (n.d.)		X		
Ford Motor Co.	X		X (n.d.)		X		X
General Electric Co.	X				X		X
General Motors Corp.	X			X (1997)	X		X
Georgia Pacific					X	X	
Mobil Corporation	X	X			X		X
Monsanto	X	X	X (n.d.)	X (1997)			X
Occidental Petroleum	X	X	X (1997)		X	X	
Olin Corp.	X	X	X (1997)			X	
Phillips Petroleum Co.	X	X			X	X	X
Procter & Gamble Co.	X	X		X (1997)	X	X	X
Rhone Poulenc	X	X		X (1995)			X

(Continued)

Table 5.1. *Continued*

Company	EBH Client[a]	CMA[b]	Responsible Care[c]	BCSD[d]	NEDA[e]	GEMI[f]	Charter Signatory[g]
Shell International		X		X (1995)	X		X
Southern Power Co.	X					X	
Texaco Inc.	X		X	X (1995)	X		X
3M		X		X (1992)			X
Union Carbide Corp.	X	X	X		X	X	X
Weyerhaeuser Company	X			X (1995)	X	X	

[a] Clients of E. Bruce Harrison, either direct clients or via a coalition managed by the E. Bruce Harrison Company public relations firm. For sources, see Appendix 2.

[b] Member of Chemical Manufacturing Association (CMA), the largest American trade association for the industry. Sources: CMA Minutes of Meetings (a) Environmental Management Committee, 21 May 1986; 26 June 1986; 6 August 1986; (b) Federal Government Relations Committee, 20 October 1995; 16 November 1995; 1 December 1995. See Papers of the CMA, Chemical Industry Archives, University of California-San Francisco Library.

[c] Responsible Care is the international chemical industry's voluntary environmental compliance program, developed by the CMA in 1988. Dates in brackets indicate the date the company became a member, if available.

[d] The Business Council for Sustainable Development (BCSD) was formed in 1992 for the United Nations Conference on Environment and Development (UNCED) in Rio de Janeiro. In 1995, the BCSD merged with the World Industry Council on the Environment (WICE) and became the World Business Council for Sustainable Development (WBCSD). Source of company membership: Lloyd Timberlake, "Catalyzing Change: A Short History of the WBCSD," Geneva: WBCSD, 2006, 74–76; Stephan Schmidheiny, Rodney Chase and Livio DeSimone, "Signals of Change: Business Progress toward Sustainable Development," Geneva: WBCSD, 1997, 5.

[e] The National Environmental Development Association (NEDA) is an umbrella coalition created and maintained by the E. Bruce Harrison Company public relations firm. See chapter 3.

[f] The Global Environmental Management Initiative (GEMI) was developed and maintained by staff at the E. Bruce Harrison Company public relations firm. It was also housed within the PR firm (the GEMI street address was that of the Harrison firm). Source of company membership: Global Environmental Management Initiative (GEMI). *Total Quality Environmental Management: The Primer.* Washington, DC: GEMI, 1993; "Value to Business: Global Environmental Management Initiative," Washington, DC: GEMI, November 1998; and Susan Moore, "Environmental Improvement through Business Incentives," Report prepared by GEMI Incentives Task Force, Washington, DC: GEMI, 1999.

[g] The Business Charter for Sustainable Development was crafted by the International Chamber of Commerce in 1991 ahead of the WICEM II meeting in Rotterdam. The charter, signed by nearly 200 companies, was then presented at UNCED as a sign of companies' voluntary commitments to environmental protections. Source for signatories to the Charter: International Chamber of Commerce, "The Business Charter for Sustainable Development: Supporting Companies and National Business Organizations: List as of 31 March 1991." Paris: International Chamber of Commerce.

Communication in the Age of Sustainable Development," Harrison elaborated his concept:

> The Rio meeting clearly foreshadowed the stormy process by which sustainable development will evolve from a mantra to real policies forged by hundreds of parties with conflicting aims and motives. In the midst of the tempest, it will fall to communicators to build support for a vision of our planetary future that can reconcile and accommodate greening and growth. This is where sustainable communication comes in: *it will illuminate the road to sustainable development.*[49]

"Sustainable communication," for Harrison, was a form of environmental risk management rooted in "soft" approaches to environmentalism. By promoting voluntary environmental compliance programs, industry benchmarking, strategic alliances with environmental organizations, and proactive disclosure, all in terms of "sustainable communication," Harrison could participate in the control of sustainability debates and leverage his expertise as the prime mover of such commitments. EnviroComm would promote the value of sustainability through the professional tools and techniques of public relations that Harrison had helped to develop. Unlike the short-term, crisis-response mode of most corporate public relations at that time, Harrison defined sustainable communication as a process of continuous engagement:

> Environmental communication should be used to help integrate corporate environmental goals, the ever-growing body of global regulatory requirements and the expectations of critical publics. In fact, communication devices can and should be used in strategic business planning to anticipate expectations and requirements, deal with critical negative opinion, and create useful partnerships.[50]

By 1994, Harrison had changed the name of AEF/Harrison International to EnviroComm and had established a network of ten PR firms in ten European countries. Harrison's choice of European PR firms to join the EnviroComm network was motivated by these firms' prior experience working with clients in the tobacco industry (Harrison himself had worked extensively with R. J. Reynolds).[51] The network operated on a franchise model. Each firm paid Harrison an annual membership fee of US$10,000 and was additionally required to "spend at least US$50,000 on advertising for EnviroComm in media with substantial readership among business and governmental executives" in their country.[52] Franchisees were also expected to use the name and resources of EnviroComm in their marketing and the title of their office locations. Public relations firms in the EnviroComm network in March 1995 included:

E. Bruce Harrison Company (USA)
Beau Fixe (France)
Bikker Communicatie (Netherlands)
EnviroComm Europe [Secretariat] (Brussels)
GörmanGruppen (Sweden)
Mistral (UK)
Promotiva (Finland)
Plaza de las Cortes (Spain)
GAIA Srl (Italy)
Trevor Russel Communications (Switzerland)
Interel (Belgium)
Arvizu, S. A. de CV (Mexico)
ITESM (Mexico)[53]

European PR firms were attracted to the network for a variety of reasons. A combination of environmental scandals and disasters, as discussed above, had substantially decreased public trust in corporate affairs while also strengthening calls for governmental regulation. EnviroComm promised to "certify" network members as having specialized knowledge in environmental communication and the ability to impose standards on client organizations that would not require government oversight. Second, EnviroComm network members, all independent firms, hoped to rise in the international rankings of PR billings to attract clients. Although some of the firms joining the network were top-ranked nationally, they could not compete with the massive multinational PR firms such as Hill & Knowlton or Burson-Marsteller. Presenting themselves as a group allowed the network members to combine resources for ranking purposes.

The Promotion of Green Communication

To implement the EnviroComm vision, its members engaged in a series of information-sharing, capacity-building, and rule-setting practices that would further cement the network's reputation and ensure the success of a green communication objective: integrating environmental concerns into the corporate business model through voluntary initiatives and self-regulation mechanisms that would anticipate and stave off global regulatory requirements.

Harrison met directly with network members two or three times a year. At these meetings, EnviroComm's members shared best practices, discussed political challenges, and debated future courses of action for the network. EnviroComm's core team based in Brussels and Washington produced regular bulletins, training manuals, and guidelines, and circulated them among member firms. An important piece of the EnviroComm system was the Responsible Care Program. Responsible Care is a voluntary industrial compliance program developed in 1989 by the Chemical Manufacturers Association (CMA) in response to public outcry after "the world's worst industrial disaster," a gas leak from the Union Carbide Corporation's pesticide plant in Bhopal, India, in 1984. The company, and the industry at large, was subject to strict regulation in the context of the rising right-to-know movement.[54] Developed while Union Carbide CEO Robert Kennedy was president of the CMA, Responsible Care was a concerted attempt to improve the reputation of the company and the industry as a whole. Indeed, the adoption of Responsible Care by the CMA itself was part of the trade association's own effort to become "a public relations promoter and private performance regulator."[55] As Kim Fortun writes, "Responsible Care established the institutional structures through which public concern about chemicals would be articulated."[56] While aspects of the program are dedicated to managing risk, a central function of the program is to manage information, ensuring that the industry maintained a hold over how it was represented in public forums.[57]

Since Harrison had begun his public relations career with the CMA and remained, in the early 1990s, a regular attendee at CMA events and panel meetings, at certain points conducting legislative monitoring for the association; and since Union Carbide was a client of E. Bruce Harrison, it was not surprising that Responsible Care was one of the tools proposed by EnviroComm to its European clients. EnviroComm advocated a multi-level communications program to implement Responsible Care: operational guidelines and program recommendations to plant managers and company divisions; employee activities such as lunch-hour events where Responsible Care films were shown to educate staff; and community relations. Here EnviroComm proposed that companies create community advisory groups to hold meetings "to inform neighbors about environmental advances at individual plants." They proposed that information about the Responsible Care program and other environmental measures "be distributed to local schools for classroom use"; and they proposed that "letters . . . be sent to leaders of the community inviting them to share your company's Responsible Care commitment and appeal to leaders to adopt similar principles in the locality."[58]

Following in the footsteps of Ivy Lee, John W. Hill, and the many other PR professionals who throughout the twentieth century shaped their clients' environmental response, EnviroComm introduced industrial environmental concerns as first and foremost problems of information, which EnviroComm experts could solve with their communication skills. The bulletins and guidelines produced by EnviroComm presented industry leaders as the creators and shapers of environmental information rather than its recipients. For example, describing the need for environmental reputation benchmarking, the EnviroComm guidelines explain:

> In the world today, billions of dollars have been invested in raising the level of environmental performance within the private sector. Billions more will continue to be invested. Yet, missing from this very expensive equation is an agreed-upon method for judging what level of environmental performance is acceptable, and who gets to define environmental performance.[59]

For the PR counselors involved in the network, EnviroComm was a vital source of knowledge about corporate environmental issues. EnviroComm's members and clients were impressed by EnviroComm's US standing, which helped to create a source of competitive advantage for these firms in securing clients. As one PR counselor in the network explained, "[EnviroComm] gave us a listening ear for environmental aspects. . . . [Clients] recognize us as a consultancy that was having this knowledge and experience in the field." A second counselor interviewed said, "[EnviroComm] allowed us to think about environmental issues and specialize and focus on the environment. . . . [I]t gave us special notoriety. We're not just . . . in Brussels or Belgium but we have this huge network. If you have an [environmental] issue in Spain, well, we can help you. So, we thought of it as an official trump card." The network invested clients with greater visibility. A third counselor noted,

> The field of environmental communication was very incipient in Spain when we joined the network. It was too novel. Our team felt a bit lost at

EnviroComm's kick-off meeting in Rome. We received the "decalogue" of environmental communication.... Looking back, I believe it was beneficial for Spain to become part of a global network, share ideas, and learn from countries like Germany that were more advanced in this field.

EnviroComm drew further legitimacy from Harrison's role at UNCED and adhered to the principles outlined in Agenda 21, most notably in Chapter 30: Strengthening the Role of Business and Industry. This chapter highlights the importance for business and industry in "recogniz[ing] environmental management as among the highest corporate priorities and as a key determinant to sustainable development," through voluntary initiatives and self-regulation. Examples included the implementation of Responsible Care and "product stewardship policies and programmes, fostering openness and dialogue with employees and the public and carrying out environmental audits and assessments of compliance."[60]

EnviroComm also looked to promote capacity building among its members. In a series of "issues briefs," EnviroComm circulated details of ongoing environmental standards processes in Europe, with a focus on planned European eco-management and auditing programs. EnviroComm members were encouraged to develop knowledge and experience in international business certification schemes, such as ISO 9000 quality standards, to assist their clients in gaining accreditation for environmental management systems (EMS). While technically "rules for the 'impartiality' of the EMS certifiers are likely to state that EMS certifiers cannot be engaged in activities including environmental consulting for the target company," in practice "this precept is likely to be held in abeyance for two years to encourage transfer of expertise.... when the field will be relatively small and sufficient control on a case-by-case basis can be exercised."[61]

Finally, EnviroComm circulated issue studies that would help its members advise their clients on environmental risk management. Issues covered included the conversion of "brownfields" into serviceable properties while managing concerns from potentially disenfranchised local residents; introduction of reputation management programs among investors, employees, and publics as environmental issues were translated into health concerns; and management of communications to de-escalate crisis situations.

EnviroComm was a pioneering network of communicators that defined and positioned "environmental communication" as an essential tool for the emerging field of environmental management at a key historical juncture.

EnviroComm functioned as an epistemic community in its ability to create and disseminate expertise and information, establish shared meaning systems and practices, and offer regular interaction with a range of relevant actors through private meetings and public events. Through the ongoing influence of E. Bruce Harrison and his associate members, as well as their prior experience working in contentious industries in the United States, the network gained authority among its private and public sector clients across Europe and Mexico. EnviroComm was able to embed the concept of "sustainable communication" in international corporate approaches to environmental management across an unprecedented geographic range. It diffused an American brand of corporate governance that promotes voluntary compliance programs and self-regulation rather than submitting to federal and state regulatory controls.

Situating communication as the locus of sustainability allowed EnviroComm to deflect attention away from the actual requirements of environmental sustainability, such as preventing natural resource depletion, limiting energy and water consumption, or reducing waste. At the same time, PR wielded power through a "subterranean politics" in which rankings, standards, and codes of conduct contribute to making environmentalism "observable, comparable and governable" across industries and territories.[62] This was a highly managed, information-based style of environmentalism that spoke to the bureaucratic norms and cultures of government agencies and corporate leaders, further minimizing the impact of civil society norms of publicity. The global, information-based, and future-oriented quality of the data further enforced its legitimacy. The word "global" here reflects the dominant paradigm of globalization in this time period, better understood as American imperialism and the vast expansion of multinational corporations seeking favorable trade and regulatory contexts for an American style of operations.

The export of "sustainable communication" from the E. Bruce Harrison Co. into international public relations firms helped to instill a specifically American understanding of environmentalism in international public and political decision-making arenas. The surprisingly similar environmental attitudes and behaviors of international firms, expressed as a compromise between economic growth and environmental protection, can be at least partly explained by the cultural discipline imposed by American public relations.

6

The Climate of Publicity

Climate Advocates and the Compromise of PR

> What's really the essence of the story is the emotional message. And
> the emotional message is, sometimes David wins, right? It may not
> be likely, but sometimes David beats Goliath. . . . And so how do we
> as environmentalists—how do we do a better job?
>
> —James Hoggan

James Hoggan may be the best-known public relations consultant in Canada.
Indeed, he is one of the few PR professionals with any name recognition at
all. One reason for this notoriety is that he is also a staunch environmental
advocate. In addition to running a crisis communications consultancy, he is
a co-founder of DeSmog Blog, a research and information center that since
2006 has accumulated expertise and materials devoted to exposing misinformation around global warming. The center, essentially a website with a small
staff of researchers and reporters, has become a valuable source of investigative material for media outlets covering climate change issues and legal
battles over climate policy. It has also become a thorn in the side of many of
the worst environmental offenders: tar sands producers or refineries, major
oil and gas companies, and other extractive industry players who attempt
to hide an inconvenient truth: that the outsized environmental hazards they
create are a constant byproduct of their normal operations.

In this dual role, Hoggan is also devoted to making space for public relations as a legitimate practice for environmentalists. His books expand the
boundaries of strategic communication, drawing on insights from thinkers
as diverse as the French philosopher Bruno Latour, American social justice
advocate Marshall Ganz, and the Dalai Lama to make the case for PR as an
open-ended and democratic process by which consensus can be achieved:

> Climate change is really an amazing gift, in a way, of a problem, because
> it is a problem unlike most problems. You are not going to fix it on your
> own. You and your friends are not going to fix it on your own. You and your
> friends and all their friends are not going to fix it. This is really a problem

A Strategic Nature. Melissa Aronczyk and Maria I. Espinoza, Oxford University Press. © Oxford University Press 2022.
DOI: 10.1093/oso/9780190055349.003.0007

of the commons. The adversarial system has its limits with that kind of problem. Just looking at it from the point of view of communication—you need to be able to have conversations where people don't just dismiss you as one of them or one of the other side.[1]

Inherent in this framing is a vision of PR as a value-neutral practice whose role is to fairly represent the viewpoints of its clientele in a range of public environments. In this optic, the problem of PR is not the PR itself; it is the ethical stance of its practitioners, its clients, or its desired outcomes. We can see this framing in DeSmog Blog's mission statement on its website:

> Using tricks and stunts that unsavory PR firms invented for the tobacco lobby, energy-industry contrarians are trying to confuse the public, to fore-stall individual and political actions that might cut into exorbitant coal, oil and gas industry profits. DeSmog is here to cry foul — to shine the light on techniques and tactics that reflect badly on the PR industry and are, ulti-mately, bad for the planet.

Hoggan and DeSmog Blog are among several recent efforts to undo the boundary work that has historically separated industrial PR from issue-oriented advocacy. On the one hand, we can see this as PR for PR—part of a broader attempt to claim legitimacy for the profession as a whole. Along with public relations scholarship, academic departments, and professional associations, Hoggan promotes an expansive understanding of public relations as a deliberative force, capable of generating rational and inclusive debates with an eye to the common good.

On the other, this approach is intended to acknowledge the "realities" of the modern communication environment, in which PR is an ideal mechanism to bring climate change to the fore of contemporary popular and political debates. In its mission to engage multiple publics, construct compelling information and influence campaigns, and deploy various strategies and styles of communication (rational, emotional, rhetorical, storytelling, testimony, etc.) to achieve its ends, PR appears to offer a superior set of methods to apprehend the "super-wicked problem" of climate change in an equally "super-wicked" climate of publicity.[2]

Promoting public relations as technology rather than as ideology strips the practice of its industrial and market orientations in an effort to widen the terrain on which it can be applied. Political communication scholar Jarol Manheim argues that strategic political communication methods are at the disposal of all kinds of actors, from policy advocates to social movements and insurgent groups and from corporations to governments, helping them "get what they want" in public and political affairs.[3] This chapter tests that premise by examining the use of public relations by environmental advocates to promote action on climate change. If previous chapters considered the practice of public relations at the level of cultural legitimacy and political infrastructure, this chapter takes us into the motives and justifications of public relations practitioners. We draw on interviews with communication strategists, academic researchers, and self-styled consultants in a range of organizations: nonprofits, nongovernmental organizations and quasi-NGOs, sustainability centers, social change and public interest groups, all dedicated to the promotion of action around climate change.[4] We asked three broad questions: How do environmentalists make sense of PR in relation to their profession and their professional selves? What kinds of "publics" and "relations" do environmental advocates develop using PR, and how does this affect their goals? Finally, what climate of publicity is created in this process? That is, what does this knowledge do for such an intransigent client as "the climate" in the public sphere?

In asking these questions we aim to further trace the epistemic dimensions of PR itself. *What does public relations "know" about environmental advocacy?* That may seem like a strange question, especially in light of the tendency we mark above, to treat PR as unaffiliated intermediary for already existing knowledge. But as we have seen throughout this book, producing public relations as technology relies on forgetting the contexts that shaped it—most centrally, the public's burgeoning consciousness of the natural environment and its awareness of environmental damage as a public problem. In showing how environmentalists have adopted the logic and practice of PR to build their advocacy claims, this chapter takes aim at the notion of PR as a value-free enterprise. As we saw in chapter 5, PR is steeped in the values of its object of promotion. The ostensible subordinate positioning suggested by the promoter-client relationship masks the agential and creative function of PR professionals as well as the material and ideational contexts of PR's elaboration on the American political landscape. Here, we see how the techniques of publicity made available within the bounds of PR narrow the scope of the identities, audiences, messages, and meanings that can be promoted.

Before we can assess the role of climate action PR in the social drama of climate change, we need to understand the origins of modern public

relations thinking about environmental attitudes and its impact on subsequent notions of public awareness and action around the environment.[5] To
demonstrate the power of public relations as a system of information management around environmental issues, we turn to one of its most influential
proponents: the academic and PR theorist James E. Grunig and his authoritative concept of situational publics.

James E. Grunig and the Situational Theory of Publics

In 1977, James E. Grunig was commissioned by the American Foundation for
Public Relations Research and Education to review the state of the field of environmental public relations. Grunig was at this time an associate professor
in the journalism department at the University of Maryland, having earned
a bachelor's degree in agricultural journalism from Iowa State University in
1964, a master's in agricultural economics from the University of Wisconsin
in 1966, and a PhD in mass communications in 1968. A science writer and
student journalist in the early 1960s, by the early 1970s he had begun to
make a name for himself as a scholar of public relations. In this era, most PR
departments were housed inside schools of journalism. Theoretical or scientific programs of PR were rare. Grunig had bigger plans for PR: he worked at
the level of systems and models. He would eventually go on to become one of
the leading theorists of the profession, developing principles of communication behavior as well as comprehensive and programmatic benchmarks for
public relations administration across organizations.[6]

For the American Foundation for Public Relations, Grunig's review took
stock of the latest research on mass communication, ecological concerns,
and attitude formation. As he noted, this research contained two guiding
assumptions about publics and environmental problems. The first assumption was that if the public develops the "proper attitude" about the environment, it will behave in a way that helps preserve the environment. The second
assumption was that "if the environment is covered sufficiently, and 'properly,' in the mass media . . . the public will develop proper attitudes toward
the environment."[7] These assumptions worked on further assumptions: that
speaking to more people was better than speaking to fewer people; that environmental problems should be framed to appeal to as broad a public as possible; and that the more people know, the more likely they are to form strong
attitudes toward the environment and alter their behavior.

Grunig suspected that something was wrong with that set of beliefs. The research treated information as an automatic catalyst, drawing a throughline from information seeking to knowledge to attitude change to altered behavior. It also construed mass audiences as fairly undifferentiated in their attitudes and behaviors: as long as the information was sound, people's knowledge, attitudes, and behaviors would change. Neither idea made sense. For one thing, by the 1970s, communication researchers had moved beyond wartime theories of propaganda that treated people as an undifferentiated mass target that could be programmed by information. There was also no automatic link between information and knowledge, or between knowledge and attitude or behavior change. People had different reasons to develop "communication behaviors," as Grunig called them: reasons for why and how they collect and process information. They may collect information to reaffirm existing beliefs. Or they may collect enough information to be aware of a problem but see it as unimportant to their lives and do nothing. Perhaps most critically, it was unclear when a problem became a problem: what was the tipping point at which the environment surfaced as a problem requiring public attention? And what was the "proper attitude" needed to resolve it?

Grunig was compelled by an idea developed by his colleague Keith R. Stamm a few years earlier. In the late 1960s, Stamm, a young journalism professor from Wisconsin, had begun researching how people developed an "ecological conscience." The use of concepts such as "knowledge," "opinion," and "attitude," he argued, "does not often capture what is significant and revolutionary about the ecological perspective: that it is a different way of 'picturing' the phenomena of our environment."[8]

To study how people developed an ecological perspective, Stamm focused on how they made sense of the concept of environmental conservation. Working on the premise that people develop attitudes toward conservation on the basis of whether they perceive the environment as a scarce natural resource, Stamm created a survey about current conservation problems in his state: low trout populations, flooding, erosion, pesticides, and water shortage, among others. For each question respondents chose one of two possible responses: "reversal of trends" (given the declining availability of natural resources, we ought to work to reverse this trend by preserving and protecting the environment) or "functional substitution" (as environmental resources are depleted, we can substitute another resource, relying on scientific and technological developments).

What Stamm found was essentially a model of conditional altruism. When people judged that the environmental situation posed a problem to them personally, they advocated the reversal of trends. But when the situation seemed unproblematic, or abstract, respondents tended to advocate functional substitution of perhaps yet to be developed resources. The factors that Stamm thought would matter—participation in outdoor activities like hunting, hiking, or fishing; membership in conservation groups; or consumption of environmental media—had almost no effect on shifting respondents' orientation toward environmental scarcity.[9]

Grunig felt Stamm had uncovered a principle of human nature, and he applied this principle to his work on public environmental problems. The people who seek knowledge and develop attitudes and behaviors toward protecting the environment, he argued, have to be "personally involved in situations where environmental problems occur."[10] In order to find or process information, formulate opinions, and act according to that information and those opinions, people have to be shown how the environment matters to *them*, not as a member of a broad public but as a self-interested individual. Put bluntly, the environment only became a problem when you cared about it.[11]

Grunig called this a "situational theory" of public behavior. He reasoned that members of a public would "become active," that is, want to communicate about a problem, if four criteria were in place. People had to (a) recognize the problem as a problem ("that something is lacking in a situation so that [they stop] to think about it"); (b) interpret the constraints (the degree to which they feel they have "freedom of choice" to resolve the problem or are hampered by larger and uncontrollable social, economic, political, or physical forces); (c) assess their level of involvement (their ability to imagine themselves as part of the problem); and (d) develop a referent criterion, a plan for problem-solving. Only if these four criteria are present, Grunig argued, will people seek information, gain knowledge, and be driven to action.[12]

With funding from the National Wildlife Federation, Grunig set out to test this theory.[13] Assembling a purposive sample of urban and rural residents, college students, and environmental scientists, Grunig prepared a survey to gather views on eight environmental issues: air pollution, the extinction of whales, the energy shortage, strip mining (asked of both rural and urban residents); superhighways, disposable cans and bottles, water pollution, and oil spills (these last four asked only of urban residents); dams and flood control projects, effect of pesticides on wildlife, fertilizer run-off in lakes and streams, and nuclear power plants (these last four asked only of rural residents). Each respondent was asked the following questions:

1. Do you stop to think about this problem often, sometimes, rarely, or never?
2. Do you see a strong, moderate, weak, or no connection between yourself and this problem?
3. Could you do a great deal, something, very little, or nothing personally to affect the way these issues are handled?
4. Do you know a solution to this problem?[14]

Because the questions were organized around self-interest, the answers seemed to prove the theory right. There were whale publics and super-highway publics and fertilizer publics, but no general environmental publics across all issues. Issues such as the energy shortage and air pollution did activate a broad swath of people but not a uniform set of responses: some desired active levels of involvement in the issue while others were passive. Some perceived high barriers to action while others saw few constraints. "There is no single 'public opinion,' about the environment or about all environmental issues," Grunig wrote in a later assessment of the study. Journalistic coverage, political decisions, or immediate conflicts may cause attention to rise, but these create contingent and particularistic commitments. "What waxes and wanes is not so much opinion as it is the number and level of activity of publics. Issues bring about publics, and publics come and go as events and personalities change and create issues."[15]

The National Wildlife Federation had funded Grunig to find out what might incentivize members of the public to join their organization. But Grunig had found something bigger: which conditions *create* publics who recognize a problem and want to find a solution. Grunig developed an evolutionary typology to accommodate these different levels of perception and drive to participation: *latent publics* are made up of those who do not see the situation as problematic, *aware publics* do recognize there is a problem, and *active publics* organize and discuss in order to do something about the situation.[16]

Grunig saw this typology as directly linked to John Dewey. Dewey, Grunig argued, had established "three conditions for the existence of a public": "A public arises when a group of people (1) faces a similar indeterminant situation, (2) recognizes what is indeterminate in that situation, and (3) organizes to do something about the problem."[17]

But Grunig had made an interpretive error—or perhaps, a self-interested slippage. He had identified the problem, not the solution, posed by Dewey. Dewey's concern was the "eclipse" of the public. In the context of expanding industrial, political, and technological complexity in a modern society, Dewey worried, how can a democratic public recognize and assert itself? How can it take the measure of those problems of greatest concern to the ongoing health and well-being of a society and devise a means to actively and continuously participate in shaping the definition and direction of these problems?

In fact, the situational public Grunig conceptualized was the opposite of the great community to which Dewey aspired: a solidaristic, authoritative, and interconnected public whose common interests are consciously sustained. As Dewey wrote:

> The ramification of the issues before the public is so wide and intricate, the technical matters involved are so many and so shifting, that the public cannot for any length of time identify and hold itself. It is not that there is no public, no large body of persons having a common interest in the consequences of social transactions. There is too much public, a public too diffused and scattered and too intricate in composition. And there are too many publics, for conjoint actions which have indirect, serious and enduring consequences are multitudinous beyond comparison, and each one of them crosses the others and generates its own group of persons especially affected with little to hold these different publics together in an integrated whole.[18]

What Grunig was really doing became clear over the next several years, as the theory of situational publics took hold and spread throughout the academic and professional spheres of public relations.[19] On the one hand, like the professional communicators who came before him, he was finding ways to segregate publics in order to better manage them. Also like them, he aligned his concept of the public with the principles of American democracy. Issue groups (Grunig's preferred term for interest groups) were part of the check on bureaucratic systems of power that ordinary people could provide.[20]

On the other hand, as he wrote in a 1989 report, Grunig was really after those publics who posed the greatest concern to industrial public relations managers:

Organizations need public relations because their behaviors create issues that create publics, which may evolve into activist groups that threaten the autonomy of organizations. . . . Environmental activist groups, obviously, have played a major role in formulating environmental policy, in challenging behaviors of corporations and other organizations that affect the environment, and in holding "public opinions" and influencing the opinions of environmental publics that have not yet reached the stage of activism.[21]

If Grunig's situational theory of publics emerged out of the nexus of concepts of information, environment, and the public, it was codified as a communications strategy to contain and discipline all three concepts.

The theory of situational publics might be less relevant to this story if it were not so strongly embedded in the theory and practice of public relations today. Whether in support or in opposition, scholars and practitioners dealing with public relations must pass through the "Grunigian paradigm" with its particularistic and context-specific vision of issue publics. Its origins in environmental problems are a non-negligible piece of the legitimation puzzle.

The question now is how this legacy plays out among climate advocates. Our hypothesis is not hard to discern: when activists use PR, they import its values as well. This affects how they represent who they are, what they do, and how they represent their "client"—in this case, the climate. The power of public relations is partly rooted in its ability to sow compromise and foster consensus, as we have seen in earlier chapters. This compromise function can also work recursively; that is, it can act back on climate advocates, who may see the path for climate action as one of acceptance or accommodation of existing cultural and political structures and attempt to work through and with them rather than against them. The outcome, as we shall see, is a campaign for climate action that leaves untouched the environment as an object of publicity.[22]

Public Relations as a Technology of Legitimacy for Climate Advocates

In chapter 4 we encountered a new style of issue advocacy: the voices of socially minded citizens organized around the promotion of the public interest.

Mobilizing on behalf of social causes and using the force of the law and emerging regulatory regimes, these advocates helped create the environment as a national public and political problem.

The formalization and professionalization of this style of advocacy over time was and remains a matter of some contestation. For some, real social change is a strategy "from below," not imposed "from above" by existing institutions and their professionals. In this view, civil society reform is antithetical to professionalization, with its elite orientation and establishment-rooted structural norms. The "compromise of liberal environmentalism," as political scientist Steven Bernstein characterizes it, is precisely its institutionalization and politicization, leading to a mainstream version of advocacy that is integrated into the prevailing economic order.[23] The range of critiques of environmental organizations gives some sense of the intellectual and practical disorder facing these groups.

For critics like Christopher Bosso, the emergence in the 1970s of environmental advocacy organizations like the Environmental Defense Fund and the National Resources Defense Council, Friends of the Earth, and Environmental Action signaled the abandonment of environmentalists' grassroots origins.[24] The new organizations were too "inside the Beltway," too focused on influencing policy agendas and holding elected officials accountable while ceding territory in the "ground game." What Michael Shellenberger and Ted Nordhaus term the "death of environmentalism" referred precisely to this narrowing of environmentalism as one special interest among many, with limited constituencies or allies and a dwindling ability to represent climate change as a problem of our collective future on this planet.[25] Older national environmental groups like the National Parks Conservation Association, the Sierra Club, or the Audubon Society were taken to task for a lack of racial, gender, and class diversity, which overdetermined their national membership and hampered their framing of a properly "public" and national (let alone transnational) interest. Those groups that were determined to remain organized at the grassroots were criticized for their too "gentlemanly" approach to the hard-nosed maneuvers within the power centers of conventional politics.[26]

These legacies both structure and trouble contemporary proponents of climate action. Today's environmental advocates are caught between a perceived need to look, act, and sound like the mainstream outfits with which they interact and a wish to maintain the outsider identity that characterizes their original visions of a more just and responsible world. More difficult still, making climate change a matter of public urgency and action seems to require the ability to advocate "from below" while maintaining a struc-ture "from above" to access critical resources. In the first instance, there-fore, adopting the trappings of public relations may seem to offer a middle ground: a familiar repertoire of publicity strategies available for uptake to promote social change. Indeed, for some activists, PR is a necessary tool in the cultural toolkit of today's information and influence environment. Parth, the communications director of a quasi-NGO focused on sustainability is-sues, explains:

> I think there was for a long time, and maybe still is in certain realms, a sense amongst communications teams and NGOs that we couldn't use the tools that others used to tell our story. It wouldn't be right. It wouldn't have in-tegrity. "That's advertising and marketing; we don't do that." And I think that there's been a movement. You know, I think a lot of organizations are actually realizing that you've got to use all the tools in communications to move people.

Some respondents said PR was useful to address a double problem: over-coming the difficulties of gaining attention and influence in a fragmented and polarized media space and adequately communicating the overwhelming complexity of climate change. Rory, the director of a research center focused on climate change behaviors, says:

> To be fair, I don't think we've ever seen a challenge like this before. I mean a lot of people bang their head against the wall and say, how come we haven't cracked this yet? Well, this is big. This is tough. Why was it so hard? I think we were using the wrong messages, the wrong messengers. It's a really hard thing to figure out how to communicate.

Finding the right messages and the right messengers involves a delicate dance: aligning their task with established systems of persuasion, some borrow expertise and techniques from political campaigns or commercial marketing and advertising. One climate communicator's firm hired staff that had worked on electoral campaigns for Barack Obama, Hillary Clinton, and Bernie Sanders. Mark described a multi-stakeholder PR campaign his international NGO had helped develop to seed public awareness of climate change ahead of the 2009 United Nations Climate Change Conference in

Copenhagen. From a publicity perspective, he explained, the campaign had been effective:

> Despite the fact that the policy outcomes of the Copenhagen summit itself fell short, I think the campaign was tremendously successful, because what you couldn't say anymore was that the public is unaware of the issue, that it is not registering in the public domain. I mean, before, climate change was this kind of obscure British disease that some strange people talked about in the margins of society.

For these advocates, public relations is an intermediary, or mode of translation, between them and their perceived opponents. PR lets them speak the language of their interlocutors, showing that they "get" an opposing side's views and can relate to their concerns. Ricky, director of communication at a large US-based environmental organization, framed his organization's strategic communications approach in terms of forging relationships across seemingly impassable boundaries, such as the one separating Democrats (blue) from Republicans (red):

> To be able to go into Wyoming or Utah or Colorado—which is purple, but only because there's a lot of red and a lot of blue . . . and be able to have an authentic conversation in their own terms with people who aren't necessarily opposed to oil and gas development, people whose bread is buttered that way . . .

This strategy, of pursuing conversations with groups who "are uncomfortable dealing with environmentalists on environmental terms," required an ongoing series of compromises. One compromise involved avoiding contentious battles, such as those waged over fracking or pipelines. The point was to find middle ground, and to do this meant identifying winnable issues, like those where a few extra votes from the West could forward the passage of climate-friendly legislation. A related compromise was to focus on particularistic interests and not core values. A decision-maker may be privately committed to environmental causes, Ricky explained, but publicly responsible for representing a constituency that is not. In these settings, some environmental organizations craft arguments to help decision-makers frame the problem in the interests of their group. The question of whether these interests join up with the larger cause of climate change was subordinated to the goal of winning the issue at hand:

The argument that got the guys on our side was not an environmental argument. There are environmental organizations—there are people within *this* environmental organization—who have an intellectually and emotionally difficult time making a sincere, non-environmental argument to build a case, to build alliances, to get the thing you want. They're like, "No, we've got to get them our way." Okay. Do *I* care why they're a yes? I don't.

This strategy may seem to play directly into Shellenberger and Nordhaus's concerns: the death of environmentalism through an over-emphasis on framing; incremental small-bore achievements over big picture ecological interconnectedness; and a play to establishment practices and institutions rather than a vision of radical alternatives. But to some interviewees, it was a sound technique to achieve legitimacy for ideas in oppositional contexts. These were not ideological compromises but strategic ones, deliberately crafted to attain a more profound set of ends. Respondents recounted infamous PR events that served as lessons in strategic influence—NASA scientist James Hansen's televised testimony before Congress in 1989; Al Gore's 2006 documentary film, *An Inconvenient Truth*; Republican Senator James Inhofe's snowball stunt on the Senate floor in 2015—claiming the power of these events to dramatically shape public opinion around global warming.[27] Public relations, as a technology of legitimacy, helped them shape media coverage and promote attention to the cause. Grant, the strategic communications director of an international environmental nonprofit, suggested that activists should do even more to gain media attention:

I mean, you look at the press coverage [on climate change] and you get those random stories on [page] B6 of the business section. Then the whole debate on the [TV] news night after night after night is coal workers. [EPA administrator Scott] Pruitt is in Hazard, Kentucky, you know, getting up there saying, "I'm bringing back coal," and there's workers on stage. Where are *our* workers on stage? Where's that? We could be doing that too. Like when solar factories open. . . . The environmental community hasn't done as much as we could to lift that up and structure that so that you have the charismatic former coal worker who now has a job doing renewables, who's speaking out and is getting on Fox and fighting these folks and is trained as a spokesperson.

Overall, interviewees saw PR as a nonessentialist, value-neutral practice, in line with James Hoggan's view. In this optic, the purpose of public relations is not to privilege any particular frame, such as green governmentality, ecological modernization, or civic environmentalism; it switches between them as needed to reach different audiences.[28] This complicates arguments about framing as a source of cultural meaning, especially those arguments that posit a hierarchy of value for certain frames and practices over others to construct norms of appropriate attitudes and behaviors. Rather, PR presents itself as a tool to capture whichever frame will work in the situation at hand for whichever public is targeted in the moment, demonstrating the contingency and short-term effectiveness of frames. At the same time, this situational approach further contributes to the mistaken perception of PR as technology and not ideology.

"Davids and Goliaths": Leveraging Legitimacy

Hoggan's reference to David and Goliath at the beginning of this chapter comes from his conversations with the social justice advocate and university lecturer Marshall Ganz, whose book, *Why David Sometimes Wins*, describes the successful effort by the United Farm Workers in the 1970s to gain collective rights to organize and negotiate contracts. The UFW was able to do this, Ganz explained, because of three "elements of strategic capacity": the movement's motivation was greater than that of its rivals; it had "better access to salient knowledge; and their deliberations became venues for learning."[29]

In Ganz's argument, "Davids" possess greater motivation and commitment to their cause. This dedication pushes them to develop creative strategies of action that can outperform material and financial resources. Thinking strategically through symbolic means, such as narratives and shared commitments, allows public issue advocates to access the emotional and moral resources required to act with agency in the face of great threats.[30]

Though our interviewees do see themselves as Davids in the ongoing "battle" to promote action on climate change, the problem they identified is not the lack of access to emotional or moral resources. Nor is the solution rooted in symbolic power: in the current context of public communication, environmental activists are not the only ones in possession of storytelling techniques. The real issue, as our interlocutors see it, is the lack of regard for *truth* in the modern political and media environment.[31] As Mark, the executive director of an international NGO alliance focused on climate action, explained:

> This kind of post-truth era that we've entered . . . telling the truth and having evidence is superfluous. It doesn't actually really matter. You can achieve things just by getting the messages right and touching people where they're at.

Environmental activists are already good at brandishing narratives expressing shared commitments; they have no shortage of emotional or moral resources. What they lack is legitimacy in the current realm of publicity. The environment is not a publicity problem, they acknowledge; it is a problem of ethical obligation and human community. But to encounter climate change today is to encounter it on fragmented terrain, where the battle for moral ground is waged on established and institutionalized platforms of public communication. Ricky, the director of communication at an international environmental advocacy organization, put it this way:

> We used to, as environmentalists, come in the door with media and least some politicians with a little bit of presumed moral high ground. And some of that's still there, but not in the way it used to be. I don't think we enjoy that presumption of truth [anymore].

Nearly all of our interviewees saw themselves through a David/Goliath lens, in which PR was a resource to "fight back" against established systems of order. In using PR as a resource, however, some climate advocates struggle to define their role in advancing the climate cause. They adopted nuanced and complex relationships to PR to make sense of their practice. Climate advocates find themselves wearing multiple hats to adapt to their audiences: activist, political strategist, communications director, nonprofit leader, media expert. They read political strategy reports and attend webinars about gaining press coverage and earned media. Some also produce this material for other climate communicators, preparing reports on digital trends or explainers with titles like "How to Talk About Climate" ("Be confident. Raise the urgency. Frame the choice between clean energy and fossil fuels").

The climate communicators we interviewed displayed different relationships to PR based on their personal and professional backgrounds.

Table 6.1. Climate PR Interview Respondents

Name (pseudonym)	Title	Organization	Founded
Grant	Strategic communications director	International (US-based) environmental nonprofit	2008
Henry	Director of communications	International environmental nonprofit	2013
Maria	Communications professor (retired)	Nonprofit trade association for public relations professionals	1948
Philippe	President	Strategic PR agency focused on environmental and social issues	2004
Bernard	Managing director	Strategic communications nonprofit focused on climate change and clean energy	2011
Isla	Consultant	Climate advocacy for nonprofits	2012
Paul	President	Private strategic communications and campaign firm for nonprofits and foundations	1990
Stephanie	Principal	Strategic communications firm focused on environmental sustainability issues	2009
Ricky	Director of communication	International (US-based) environmental advocacy organization	1967
Mark	Executive director	International NGO alliance focused on climate action	2009
Mario	Senior associate	International strategic communications consultancy focused on climate action[a]	2017
Gus	Executive director	International strategic communications consultancy focused on climate action[a]	2017
Lorne	Co-founder	Strategic research center focused on climate advocacy	2006
Anna	Director	Research institute focused on climate change	2013
Ramona	Senior vice-president	Communications and marketing firm for nonprofits[b]	1991
Juan	Senior strategist	Communications and marketing firm for nonprofits[b]	1991
Parth	Communications director	Quasi-NGO focused on sustainability issues	2011
Rory	Director	Research center focused on climate change behaviors	2005
Stephen	Co-founder	Strategic communications firm focused on progressive causes	2016
Dylan	President and founder	Private research and consulting center focused on behavioral science	2017

Note: [a] and [b] indicate that the interviewees were at the same firm.

Some had moved into climate change advocacy from consumer- or business-oriented public relations firms, seeing the move as a way to align their skills and training with their personal values. Others, who identified as activists, saw PR as a necessary evil, if not an easily accepted one. Henry, a longtime activist on behalf of environmental causes, was ambivalent about taking up a strategy position in a private PR firm:

> What I told myself when I went to work for Edelman [a large multinational public relations consultancy] was, I will never spin and I will not do defensive PR. I am not going to stand up and defend the misbehavior of a company, and if they ask me to do that, I'm out. I'm not doing it. I'm more than happy to help a company that is actually doing things that are benefiting the world. I will happily tell those stories. Whether it's environmental health, STEM education for young girls, whatever it is, I would love to tell those stories and help you come up with great strategies to tell those things, but I'm not going to dissemble—I'm not going to do that.

Some respondents did dissemble, admitting that developing narratives to align climate action with establishment motives was a source of discomfort but one that they tolerated as part of their organization's approach—or, more broadly, as part of the "reality" of climate change action, which required an "all hands on deck" perspective. This was especially apparent in cases where businesses had to be included in order for change to take place. Mario, a senior associate at an international strategic communications consultancy focused on climate action, expressed this ambivalence:

> Walmart is the biggest purchaser of solar panels in the United States, and people see that, and it's tangible. So, you know . . . a lot of Whole Foods people might not be on board, and I may be one of them. But that's a real thing that is happening, where you're seeing companies get on board. And people, sometimes they're a little uncomfortable at first, even if you still have that quick wince—but we're generally okay with it.

Another way environmental advocates made sense of their work with business outfits was to see their organization as part of a larger community of climate advocacy groups. It was acceptable for some groups to accommodate business Goliaths as long as there remained a "radical flank" of resistance.[32]

Parth, a communications director at a quasi-NGO focused on sustainability issues, explains:

> You always need the Greenpeaces because they're constantly pushing, right? And they're holding companies accountable. So we wouldn't have been successful without the Greenpeaces because we needed them to be pushing, pushing, pushing, and then the CEOs were much more receptive to *our* message.

The metaphor of David and Goliath was especially apparent in the responses of some climate activists, who described their communications practice in staunch opposition to PR. Grant, a strategic communications director at a US-based environmental nonprofit, said:

> I find activism fascinating because I think it provides all of these surprisingly powerful ways to communicate and intervene in the political system which are seen as sort of like outside or grassroots or less professional. Like not a lobbyist, an ad, and like "big budget" and "DC." You know, you work for Glover Park Group [a strategic communications and public affairs firm] and you do these things.

Reaffirming his outsider status was a source of authenticity. It also served as a marker of differentiation between his version of advocacy and that of industrial PR:

> We find it endlessly amusing how little the industry seems to understand the way that advocacy works. We have hundreds of thousands of people on our email lists. You know, we have hundreds of local groups. We have thousands of supporters. We don't need to pay them to show up at places. We don't need to have like paid provocateurs. Like, there's a real movement, you know.

Unlike PR industrialists, who engage in "grassroots for hire" practices or use PR to create "fake" (industry-sponsored) citizens' groups, Grant argued, climate advocates have "real" supporters who actually care about the problem of climate change and can be mobilized in support of climate actions.[33]

What comes to light in these examples is that the David/Goliath metaphor is itself an act of strategic communication, a politically expedient binary that emphasizes degrees of distance from the machinations of industry.[34] That this binary glosses over the complexity of organizational commitments, overlapping allegiances, and cultural diversity characteristic of public life reveals the limits strategic communication can place on popular understandings of climate change or related concepts.

It would be a mistake to characterize these responses as capitulation or resistance to big business, however. Nor are they reducible to a pattern of so-called corporatization of activism.[35] Rather, we propose to see these respondents' vocabularies of motive as first and foremost claims for status in the contemporary climate of publicity. To make an interested public in this framework is a matter of gaining legitimacy for yourself as much as for the ideals you stand for; in turn, what you communicate *about* gains credence by virtue of being publicized. In this context, PR is the source of legitimacy for both actors and issues. Making use of the repertoire of PR is also about gaining access to the networks of legitimacy that go along with it: elite policy networks, funding and board alliances, and other relational structures both external to and internal to home organizations.

The ultimate legacy of the situational theory of publics is that this process of legitimacy-making assumes that publicity—creating certain areas as matters of public concern subject to popular decision, and then creating publics who are "activated" or mobilized around these matters—is enough. But rather than promoting and sustaining common cause or solidarity through interconnectedness (either ecological or social), public relations places people and issues into the fragmented, contingent, and oppositional discourse realm characterizing democratic decision-making today.

To address the enormity and urgency of global anthropogenic climate change, all respondents expressed a desire to re-create a civic polity, mobilizing a democratic, transnational public whose force of solidarity would form the long-term general will to draw up plans of action and redress. Gus, executive director of an international strategic communications consultancy focused on climate action, asks:

How do you shape public discourse? That's the question that I think strategic communications has to ask and indeed answer. How do you create a political environment? Essentially how do you create the tide that lifts all the boats? How do you actually change the context in which those decisions are getting made rather than just the decisions themselves?

In principle, the cause transcends the will of individuals and the vagaries of public opinion. But in the details, climate advocates constructed publics as multiple and contingent, where influence was gained by appealing to self-interest and particularistic qualities. To the extent that moral obligations were offset by the realities of the deliberative systems in which legitimation takes place, public relations was the strategic tool by which climate advocates reached these various publics.[36]

One respondent explained that the focus on situational publics was a structural effect. Given the capacities of digital media and political campaigns to fragment, monitor, and target publics along carefully defined data-rich lines, there was no real alternative for climate advocates but to adopt the same approach and decentralize their outreach. By targeting "the right people in a non-unified way," the climate movement felt its efforts were gaining traction. The use of situational publics was also a result of the fragmentation of the climate movement itself. While fundamentally a science or environmental issue, climate change appears on the national political agenda within different frames: as a matter of human health, economic growth, security, or energy consumption, among others. As it moves through these frames, different publics are activated to respond.

Synthetic Narratives: Legitimacy by Proxy

To talk to their multiple publics, climate advocates use stories:

> Lean toward stories because it's basically the way people process information. You want to get people's attention, you want to win their hearts, you want to win their minds, you want to sort of pull us all together and everything? We need to become better storytellers. There is no doubt in my mind at all that a big part of all these environmental problems is the failure of advocates to really communicate well. And we need to become better storytellers. (Lorne, co-founder of strategic research center on climate advocacy)

Climate advocates also use stories as vehicles for other stories:

So a fossil fuel divestment campaign becomes an entire way to talk about everything from financial risk of investing in fossil fuel companies to the moral argument of why you shouldn't do that anymore to talking about green investments and new opportunities. (Grant, strategic communications director)

But the most valuable stories for climate advocates are those that serve as scripts for action. These are what one interviewee called "synthetic" narratives: projections of *commitments* to climate-friendly behaviors, if not actual results, in order to provoke others into making similar commitments.

Synthetic narratives are informal corollaries to the formal rule-setting of transnational climate agreements. Just as climate agreements involve promises by stakeholders to act toward predetermined objectives, such as levels of carbon dioxide or sector adoption of renewal energy, synthetic narratives are the stories that convey the power of these commitments to a broader audience.

Climate agreements such as the Paris Accord are synthetic narratives in their own right. The Paris Accord, a global climate change agreement ratified by nearly 190 countries in 2016, asks each country to outline and communicate its projected climate actions, known as Nationally Determined Contributions (NDCs). One respondent referred to this as a "ratchet" system of evidence. As one country makes a commitment to improved climate behavior, the premise is that this will ratchet up the obligation for others.

Parth, the sustainability communications director, described an initiative called RE100, a partnership of business groups committed to using entirely renewable energy sources by a projected date of 2050, to explain how synthetic narratives act as a kind of ratchet for action:

We built a menu of initiatives that were already in existence, but we thought were the most credible initiatives companies could sign up to [in order] to really take action, and that were trackable. That was the key. [We] were tracking the progress on these initiatives. So RE100 was an initiative that we kind of created, actually, which was about companies committing to going 100 percent renewable for energy. . . . What that did is it changed the game, because we were able to then go to the negotiators and say these companies have skin in the game now; they are moving, and we are tracking them. And *they* want you to do this. And we came up with a menu of eight asks that we wanted out of Paris [the Paris Accord], that business wanted out of Paris, 'cause it would be good for the planet and good for their bottom line. Because the old paradigm was, "You go first, business"; and then business was saying, "No, no, we need policy to tell us what to do." And it was this very comfortable stand-off. And what we did is we just broke that. . . . It was like, "We're already in. If you raise the level of ambition, we can raise it more." "If you raise it more, we'll raise it more." It became this very powerful dynamic. And we ended up getting all eight asks in the Paris negotiations.

Climate advocates' use of synthetic narratives is seen as especially important in the American context, where national-level political consensus on climate change is all but nonexistent. Synthetic narratives are a "bottom up" strategy that can compensate for US federal government inaction on climate policy, by incentivizing subnational stakeholders such as state-level actors or corporate CEOs to act—or promise to act—and to promote those possible actions to multiple audiences. Whether or not these promissory notes carry weight at the federal level, they form a powerful context for action in which participation is valorized. In this regard they are the ideal public relations strategic tool: they perform legitimacy for various audiences, constituting cultural evidence that is not beholden to scientific proof. Moreover, this strategy of publicity highlights the lack of participation by uncooperative actors, who can then be brought into the storyline as rogue antagonists or uncaring enemies.

Synthetic narratives can be many things; but certain manifestations of evidence are perceived as more impactful than others. Quantification of action is revered, combining the apparent objectivity of numbers and clear articulation of metrics with the superiority of a datafied template. Mario, the associate at the international strategic communication organization, explains:

Let's say if you got a million people to say, "I will not eat meat on Mondays." You could quantify what that carbon impact would be. And then, so if you could say, "All right, we've got a million people—Americans—to, you know, essentially reduce their meat consumption by 20 percent, and we have 50 states."

This kind of knowledge was seen as powerful evidence not only for policymakers but also for ordinary citizens who may be more moved by rational (numerical) arguments than by the emotional tenor of stories.

As critics have deftly noted, corporate actors make extensive use of synthetic narratives, especially their quantitative features, as part of their social responsibility initiatives.[37] Schemes of carbon accounting, environmental profit and loss statements, and less-frequent company financial reports are initiatives that are meant to demonstrate long-term commitments to structural change.

But they also demonstrate the propensity of these promises to become promotional in and of themselves. Like climate polls or carbon offsets, climate narratives promote a future in which we eventually exit our situational public mode and enter into the Great Community that John Dewey envisioned. In the present, however, they are more effective as PR for the organizations themselves.

One conclusion we can draw in this chapter is that turning the climate into a "client," no matter how pure the intentions or how morally right the motives, removes it from its physical basis, its co-location in the atmosphere, the biosphere, and the other related systems of land, oceans, and air. More consequential still, it elides the profoundly human nature of the problem of climate change. PR for the planet is a culturalist phenomenon. It considers humans as cultural creatures, whose attitudes and behaviors can be forged by frames and messages or stories that appeal to us as individuals on the basis of our self-interest. It returns us, as William Cronon put it, to "the wrong nature," one in which "too many corners of the earth become less than natural and too many other people become less than human."[38]

But this is an unfair assessment if we consider the structural limitations in which climate advocates find themselves. When PR professionals take on the climate as their "client," embracing the issue for its own sake, we might expect that it will make discourse about climate change more disciplined, more strategic, and more politically performative, and indeed this is the case. PR changes what counts as publicity. Anchored in the relative propositions of legitimacy, PR is bound by its focus on present situations and influential targets. In its bid to render climate change more visible and meaningful for media, politics and business, climate PR ignores, excludes, and silences those paradigms, plights, and constituents whose concerns are less palatable and especially less amenable to information-based resolution.

In chapter 7, we shall discover how public relations is moving further in this direction, not away from it. In the new nexus of information, environment, and publics heralded by the data economy, established systems of knowledge are subject to tournaments of value, where the prize is awarded to the information that best serves those who can harvest it, attempting to make publics even more amenable to their cause.

7

"Shared Value"

Promoting Climate Change for Data Worlds

In 2014, United Nations Global Pulse, a "data innovation hub and knowledge center" promoting public-private partnerships for development projects, launched a Data for Climate Action (D4CA) "challenge." Companies from the technology, retail, finance, and telecommunications sectors provided anonymized, aggregated datasets to teams of data scientists and researchers who used this to devise pragmatic solutions to address climate change. Inspired by France Telecom Orange's 2012 Data for Development challenge, the ostensible objective of the D4CA campaign was twofold: first, to show the public sector and the research community how private sector data can be used to achieve Sustainable Development Goals (SDGs); second, to establish a model of secure data provision to encourage multiple companies to participate with minimal risk to their proprietary data elements.

D4CA challenges (a second one was held in 2017) advance what UN Global Pulse calls "data philanthropy"—a data sharing practice by which businesses "donate" their data to serve the public good. Also called "data for good" or "data for development," the practice has gained adherents in both the private and public sectors since the concept was introduced at the World Economic Forum in Davos, Switzerland, in 2011. Data challenges such as the D4CA campaigns have captured the imagination of the private sector and the general public, singing a new song with harmonies of global participation to drown out the growing public chorus about the harms of consumer data collection in terms of privacy and security, transparency and legality, and rights and equity. Adherents point to the immense potential of big data as an information resource to help personnel and citizens respond more quickly and efficiently to urgent social problems such as humanitarian aid distribution or epidemic control, aiming to elevate the promise over the perils of personal data collection. D4CA is now considered a key area of intervention for the broader data philanthropy movement; and conversations around D4CA are taking shape in a number of contexts, from UN climate

A Strategic Nature. Melissa Aronczyk and Maria I. Espinoza, Oxford University Press. © Oxford University Press 2022.
DOI: 10.1093/oso/9780190055349.003.0008

summits to urban and regional planning events, and from business conferences and hackathons (e.g., Bloomberg's annual Data for Good Exchange) to corporate social responsibility programs (e.g., MasterCard's Center for Inclusive Growth).

The data for good formulation can be seen as a response, or counteroffensive, to the emergent regulatory oversight of national governments and international organizations over the misuse of personal data ("digital data created by and about people") and the overreach by technology companies. This regulatory impetus reached its apex with the passage in 2018 of the General Data Protection Regulation (GDPR), a law passed by the European Union to maintain the privacy and security of individuals' personal data, with sweeping impacts on organizations worldwide. Data for good adherents attempt to offset the image of unethical or uncaring data-collecting organizations perpetuated by regulatory regimes such as the GDPR, generating arguments for multiple audiences that present data expropriation and reappropriation as not only safe and just but also essential for knowledge and action around global public problems.

When engaging with private stakeholders, data philanthropy advocates present the practice as a business opportunity, generating what Michael Porter of the Harvard Business School calls "shared value," whereby social problems are made into "productivity drivers" for firms. In this framework, becoming a "data donor" is a means to maintain supplies and profits, reach new markets, and expand technical infrastructures.[1]

When oriented toward a public audience, data philanthropy frames social problems as "lack of information" problems, where big data can fill crucial gaps in knowledge, whether spatial, temporal, or demographic, and provoke more robust responses in terms of accuracy, timeliness, or adequate resources.[2] In this frame, the private sector is positioned as a critically important social actor in resolving development problems by sharing valuable data with national statistics offices, development agencies, and research centers. Even more consequential, data philanthropy is heralded as a first step toward the creation of a "data commons," a public space to house valuable social data that can be accessed by multiple actors for the good of all. Sister initiatives to D4CA, such as AI4SDG (Artificial Intelligence for Sustainable Development Goals), embrace data philanthropy as a move toward a voluntary regime of environmental governance and accountability that can engender global health, equality, and well-being outside formal regulatory structures.[3]

This logic is made palatable to end users (i.e., the individuals whose behavioral, locational, or other data have been collected into a privately owned dataset, with or without their knowledge) by appealing to the notion of mutual obligation. Individuals who opt out of a data commons are said to create both a "free rider problem" (e.g., benefiting from data-mined policy research without having to contribute their own data) and a "tragedy of the commons" since "the collective benefits derived from the data commons will rapidly degenerate if data subjects opt out to protect themselves."[4] By participating in D4CA and related challenges or initiatives, end users can work collaboratively with private companies and other "stakeholders" to promote the use of data for good.

This chapter subjects these various premises to critique through a detailed examination of the activities of UN Global Pulse and its various collaborators and advocates in the promotion of Data for Climate Action initiatives. In promoting corporate-owned big data as a solution for problems of environmental and climate destruction, UN Global Pulse and its collaborating organizations perpetuate the spirit and practice of publicity we have been examining in these pages. First, D4CA reveals the ongoing preoccupation by the private sector to maintain a positive image among its various audiences; and this preoccupation drives the development of information strategies—and the reliance on information brokers—that perpetuate problematic understandings of what it means to act together as a public. Building the D4CA campaign around shared risks and self-interested rewards, UN Global Pulse and its partner organizations imagined data as a common currency that could be transacted to create value for its participants—who were themselves imagined as "stakeholders" with much to gain or lose by their investment in the problem of climate change.

Second, D4CA is centrally about promoting the expertise of the private sector as a specialized and necessary complement to scientific findings by climate researchers. In this sense, D4CA rehearses the performative dimensions of corporate activities, whereby companies engage in "creating numbers" such as carbon markets to measure the sustainability efforts of the firm or promote environmental information systems that rely on private-sector data and infrastructure for decision-making around environmental issues.[5] In the process, these performative techniques change what counts as an environmental problem and which actors are best equipped to solve it.[6] As we saw in earlier chapters, since the 1980s, the "greening" of corporations by means of their adoption of voluntary (i.e., independently developed,

self-imposed, and non-binding) practices of environmental sustainability
have not only involved new accounting, information, and audit regimes but
have also given rise to new forms of authority that decenter government and
other public sector information and experience in favor of business expertise.[7] These forms of environmental management are typically more about
the political sustainability of corporations than about their contributions to
environmental sustainability; D4CA is no exception.[8]

A third feature of the D4CA campaign is more subtle but perhaps most
revealing for comprehending the system of public relations in the contemporary context. In their drive to create situational publics around social
problems, to broker relationships among parties that can operate in their
favor, and to decenter their role as value-laden protagonists and operate instead from the sidelines as "value-neutral" intermediaries, the UN Global
Pulse and its collaborators and adherents have effectively become public
relations agents.[9] D4CA reinforces the notion that the system of PR and its
role in managing and disciplining public information and communication
remains instrumental to the organization of modern social and political life
even in digital, self-mediated, and globally accessible information worlds.
UN Global Pulse staff don't think of themselves as PR agents, which suggests
that the PR function has in the current era been distributed or diffused into
professional identity and practice more generally. This makes sense given the
affordances and requirements of contemporary media, where image management has become paramount to personal and professional lives.

It also speaks to the ongoing nexus of information, environment, and publicity in the making of an American environmental consciousness. Making
the environment—or here, climate change—into an information problem
transforms how it is constituted as a problem. Writing about the proliferation of environmental information systems such as computer modeling and
simulation to monitor climate change, Kim Fortun argues that such systems
"structure what people see in the environment, and how they collaborate
to deal with environmental problems. . . . [T]hey are technologies designed
to produce new truths, new social relationships, new forms of political
decision-making, and ultimately, a renewed environment."[10]

As a technology of legitimacy, the information system of PR has since the
early twentieth century worked toward producing this renewed environment. Its legacy is apparent in the D4CA campaign, where it operates as a
system of power, providing access to some knowledge at the cost of other
forms of knowing, and managing risks for stakeholders while diminishing

attention to the risks to global health and well-being posed by the crisis of climate change.[11]

This chapter examines the logics by which Data for Climate Action is presented to private and public sector actors as a secure, trustworthy, and legitimate means of data collection and an opportunity to participate in responding to the climate crisis. We argue that while this campaign seeks to uphold the social value of big data by presenting it as a source of necessary knowledge to solve global public problems like climate change, its ultimate goal is to preserve the practice of corporate collection and targeting of user data and to maintain the value of these data as a private asset. As such, rather than legitimating the use of big data for climate change, we show that climate change is used to maintain the legitimacy of big data.

The chapter is organized as follows. First, we examine perspectives on the uses of data by private sector actors for environmental and climate-related response, considering how claims to use private data for public good are frequently offset by their practical limitations. We then outline our research method and data collection process; we review the conceptual origins of "data for good" and the principles by which it has been made meaningful in policy contexts. We next show how the Data for Climate Action campaign and the major players involved promoted D4CA as a safe and secure way to generate value for all participants, demonstrating the relevance of this campaign for thinking more broadly about the problems posed by "data for good" paradigms in the realm of global governance. We conclude with a discussion of the impact of D4CA and related initiatives and the implications these present for responding adequately to the enormous challenges of global climate change.

Civilizing Data: Big Data and Global Development

On 31 March 2009, amid mounting concerns in Europe about the rapid proliferation of techniques by commercial organizations to collect vast amounts of digital data about individual consumers and their online behaviors, a meeting was held in Brussels to discuss potential responses. In her keynote at the event, European Consumer Commissioner Meglena Kuneva attempted to balance consumer and regulator concerns with an acknowledgment of the economic opportunities presented by information and communication technologies:

It is precisely because we want these new opportunities to grow and evolve, that we need to promote the trust and confidence that will encourage people to participate. Internet is an advertisement supported service and the development of marketing based on profiling and personal data is what makes it go round. *Personal data is the new oil of the internet and the new currency of the digital world.*[12]

The power of this metaphor—data as oil—and its implications for big data's role in addressing global social problems have formed the basis of arguments both for and against the "data for good" paradigm.

For proponents, data are indeed the gushing resource of the digital economy, with enormous value to be derived from extraction and refinement. A primary argument along these lines comes from economic organizations such as the World Economic Forum, which argues that personal data "will emerge as a new asset class touching all aspects of society."[13] At the core of this view is the strongly held perspective that a so-called multi-stakeholder approach, by which private companies participate in the problem-solving, is essential to develop innovative responses to ongoing social problems.[14]

For critics, "data is the new oil" has a rather different meaning: the activities of corporate owners to capture and derive value from personal data is nothing less than a new phase of colonialism. Social theorists Nick Couldry and Ulises Mejias argue that data companies redefine social relations to normalize the act of digital dispossession, echoing historical appropriations of resources, territory, and personhood. They identify four discursive logics by which companies obscure their practices of personal data extraction and control. First, personal data are promoted as a vast and largely untapped natural resource whose value lies exclusively in their extraction and refinement. As such, data "are 'merely' the 'exhaust' exuded by people's lives, and so not capable of being owned by anyone."[15] Second, companies' use of consumer data is not about deprivation of ownership but "just sharing," and such a benign form of reciprocity conduces to the benefit of all.[16] Third, corporations are uniquely positioned to wield the skills and knowledge required to collect, process, and analyze such vast and complex quantities of digital data. Finally, companies espouse a rationality that "operates to position society as the natural beneficiary of corporations' extractive efforts, just as humanity was supposed to benefit from historical colonialism as a 'civilizational' project."[17]

Engin Isin and Evelyn Ruppert offer a portrait of data colonialism by attending to the complex issues arising from digital development, or

ICT4D (information and communication technologies for development).[18] Though arguments in favor of ICT4D present data extraction as a necessary complement to existing data sources such as national statistics and demographics, ICT4D often reinforces hierarchical perceptions of global regions, portraying countries that are "information poor" as beneficiaries of knowledge and insights from "information rich" sites. Moreover, though long-standing manifestations of imperial data politics such as the census or the metric system produced power arrangements between colonizers and colonized, contemporary ICT4D produce data that not only identify attributes of a population but subject them to monitoring over time and on a constant basis. This emergent "data empire" allows the Global North to set the terms of data collection and interpretation in the Global South, with dramatic implications for decision-making around development issues.[19]

Nevertheless, big data enthusiasts persist in seeing datasets as a diverse, integrated and timely source of information, one that could fill considerable gaps in global knowledge and action. In chapter 5, we discovered how public relations consultants helped to promote uses of corporate data for environmental or climate action in the late 1980s and early 1990s by introducing and circulating new standards, norms, and infrastructures of environmental responsibility. These norms were enforced via PR and business networks (such as EnviroComm or the WBCSD), auditing and certification schemes (such as Responsible Care), and managerial initiatives (such as GEMI). They served primarily to circulate the idea that the private sector harbored specialized expertise to meet global objectives of environmental sustainability and protection.[20] In the contemporary context, similar patterns emerge around the promotion of carbon markets and other business-friendly climate action initiatives. Carbon markets and related forms of climate accounting are co-constituted as authoritative by a range of like-minded actors, from global governance institutions to transnational and nongovernmental organizations.[21] These efforts require considerable promotion to maintain their legitimacy. Promotional managers are especially adept at blurring boundaries between climate knowledge and business knowledge, invoking concepts like "sustainability," "climate," and "public good" to justify business activities. We should be sensitive to ongoing efforts by promotional actors to dedifferentiate the concepts of environment and data—such as using the metaphor of the "cloud" in computing services—to elevate the symbolic implications of the management strategies themselves.[22] Taken as a whole, these initiatives can be seen as attempts to shift the needle, *not* on actual environmental problems at hand, but on the way problems are defined, managed, and evaluated.

This problematic—redefining the problem instead of making inroads to solve it—is especially complex in the realm of climate change. Climate change has been defined as a "super-wicked" problem for its unprecedented spatial and temporal challenges, its obstacles to cognitive and social judgments, and its low incentive structure for those paradoxically best placed to address it in policy settings.[23] It is partly for this reason that private-sector data processing has gained a foothold as a potential contributor to climate problems: as a new kind of environmental information system that can fill gaps in global climate data sources by providing more diverse, integrated, and timely datasets. As data researchers James Faghmous and Vipin Kumar note, current climate data sources present significant challenges for researchers. The lack of long-term data; problems of heterogeneity (i.e., having to deal with a wide array of data sources that are complementary but also possibly redundant); constantly changing observation systems; limited understandings of how data were collected and with what purpose; and limited data representation models that acknowledge the climate system as a multivariate and ever-evolving spatio-temporal network; these are some of the challenges. While big data analytics could help complement current observational, remotely sensed, and model output sources of climate data, just as with any data-driven exploration, it raises questions over sampling bias, autocorrelation, and causal inferences in predictive models. The greater risk with big data analytics is to present it as the "silver bullet" of modern research, where findings can be interpreted using a "theory-free" mindset.[24] While these methods will produce results, they will yield few insights without theory.

Promoting private sector data and its analytics as essential information resources for climate concerns is rooted in the promise of these data to both define and govern environmental problems as well as to support evidence-based decision-making. Based on the principle that all citizens, corporations, and state agents require equal access to information to judge environmental problems, the notion of environmental information systems (EIS) as sources of decision-making has become engrained among environmental justice activists, government agencies, nonprofits, and corporations.[25] The category of EIS is broad, encompassing such diverse systems as remote sensors, geographic information system mapping, and visualization; computer simulators, inventories, and databases; and environmental accounting and reporting modules. EIS have been adopted to address water quality, pollution, deforestation, environmental justice, and climate change. Scholars such as Kim Fortun have referred to the proliferation of EIS to support evidence-based environmental governance as the "informating" of environmentalism. Through various means, and in varying formats, EIS help to control what a system of environmental topics, data, and expertise consists of; and how this information is communicated to different audiences.

Recent studies on the cross-pollination of big data and environmental governance have shown how environmental activists are increasingly adopting EIS that rely on big data. Environmental activists have primarily engaged with EIS that depend on voluntary data collected through participatory citizen sensing or crowdsourcing and secondarily with data-mining projects that collect social media posts about pollution and health.[26] Initiatives like these contribute to the idea that EIS produce data that are equitable, reliable, and accurate.[27] They lend credence to the notion of partnerships between the public and the private sector to secure access to big datasets that would otherwise only be used for profit (or remain unattended). These ideas have thus taken center stage in the conversation about how to harness the "data revolution" to advance the agenda of the 2030 Sustainable Development Goals.

Research Process and Data Collection

In order to assess the relevance of UN Global Pulse and the D4CA challenges in popularizing the concepts of "data for good," "data philanthropy," and "data for climate action," we first conducted a thorough review of public documentation pertaining specifically to those terms, including news articles, technology magazines, and documents published by intergovernmental and international organizations like the United Nations and World Economic Forum (WEF). We also reviewed reports and white papers authored by collaborators of the UN Global Pulse innovation lab, such as participants in the UN World Data Forum and members of the UN Secretary-General's Data Revolution Group.

After this initial stage, we contacted a list of actors who appeared prominently in the documentation. We prepared a semi-structured interview guide designed to elicit perspectives on the emergence and development of the aforementioned concepts, especially on the "data for climate action" approach. Questions covered individual professional trajectories and engagements with the field of data for good before and after participating in the D4CA challenges and other data-for-good events as advisors, evaluators, or organizers. Interviewees were also asked to reflect on what constitutes the emerging field of data for good, the practice of data philanthropy, and initiatives like D4CA; and their perceived implications for data sharing, corporate culture, and climate governance.

After an initial round of interviews with a small pool, we adhered to a limited snowball sampling method in which interviewees were asked to recommend other data-for-good experts. We repeated our method of research, approach and interview with this secondary pool. Thirty-eight experts were contacted; nineteen were interviewed. Given the high profile of the interviewees (e.g., senior executives, founders, and CEOs of tech companies), we consider the total interview sample to be significant.

All interviewees work (or have worked) in data companies, think tanks, foundations, intergovernmental organizations, and international organizations, where they occupy roles promoting private-public cooperation to advance the achievement of the SDGs or other climate change mitigation through big data. Some were data scientists, app developers, and public relations consultants; others had a background in development, climate science, or policymaking. Interviewees landed in the field of data philanthropy from a number of paths. Some had worked or served as an advisor for the UN Secretary-General's Expert Advisory Group on the Data Revolution. A number had careers in development and climate science and had worked in different UN agencies using big data to address risk reduction, disaster management, and humanitarian response. Some had backgrounds in tech companies working as developers or communications managers. At the time we conducted the interviews, a year after the second D4CA challenge, most of the respondents occupied senior-level positions in their organizations. Their ages ranged from thirty to fifty years old.

We also participated in three data-for-good events: Bloomberg's 2017 Data for Good Exchange (#D4GX); WEF's teleconference on big data for health, "Epidemic Readiness and Trustworthy Data"; and the 2019 CIBC Analytics Day, an event by the Canadian Imperial Bank of Commerce focused on the theme of Data for Good. These participant-observation activities helped us supplement the interview data with *in situ* considerations of the organizational discourses, practices, and tensions among Data for Climate Action advocates.

Table 7.1. Data for Climate Action Interview Respondents

Interviewee	Date	Professional Affiliation	Professional Title	Relation to the data for good movement
Respondent 1	4/30/18	Center for International Earth Science Information Network (CIESIN)	Director of Customer Success & Advocacy	D4CA Strategic Advisor (2017), UN Secretary-General Expert Advisory Group on the Data Revolution (2014)
Respondent 2	5/6/18	World Economic Forum (WEF)	Data Driver Development	D4CA Evaluation Committee (2017)
Respondent 3	5/10/18	Former UN and Skoll Global Threats Fund	Program Officer	D4CA Evaluation Committee (2017)
Respondent 4	6/6/18	Former UN Global Pulse	Data Innovation Specialist	Big data and Sustainable Development Goals (SDGs)
Respondent 5	6/8/18	Big Data Research, LIRNEasia	Team Leader	D4CA Evaluation Committee (2017)
Respondent 6	6/12/18	Former UN Secretary General's Climate Change Support Team	Climate Policy Advisor	D4CA Technical Committee (2014)
Respondent 7	6/19/18	UN Global Pulse	Director	D4CA Strategic Advisor (2014 & 2017), UN Secretary-General Expert Advisory Group on the Data Revolution (2014)
Respondent 8	6/19/18	UN Global Pulse	Research Consultant	D4CA Organizer (2017)
Respondent 9	6/26/18	Pulse Lab Jakarta	Chief Technical Advisor	Big data and SDGs
Respondent 10	6/29/18	CEPEI Colombia	Data Coordinator	Big data and SDGs
Respondent 11	7/3/18	The Centre for Internet and Society (CIS)	Research Director	Big data and SDGs
Respondent 12	7/16/18	UN Population Fund (UNFPA)	Humanitarian Data and Resilience	D4CA Technical Committee (2014)
Respondent 13	7/19/18	Dalberg Data Insights	Project Manager	Big data and SDGs
Respondent 14	7/31/18	Dalberg Data Insights	Data Scientist	Big data and SDGs
Respondent 15	8/17/18	Humanitarian OpenStreetMap (HOT)	Director of Community and Partnerships	Big data and SDGs
Respondent 16	9/24/18	Crimson Hexagon	Director of Customer Success & Advocacy	D4CA Data Donor (2017)
Respondent 17	9/27/18	FSG	Co-Founder and Managing Director	Shared-value expert
Respondent 18	10/15/18	DTN	Chief Meteorological Officer	D4CA Data Donor (2017)
Respondent 19	10/17/18	Earth Networks	Chief Marketing Officer	D4CA Data Donor (2017)

"A World That Counts": Promoting Data as a Global Good

"If you can't measure it, you can't manage it." This is how Michael Bloomberg announced, via Twitter, the fifth Bloomberg Data for Good Exchange (#D4GX)—an annual event that brings together corporations, policymakers, nonprofits, charitable foundations, and researchers to explore how big data can solve the most pressing social problems of our time. The 2018 conference theme was, "*Our* Data for Good?"—reflecting on ways the private sector could deploy its data assets to develop data science projects that focus "on everyone having a stake, making it solid, fair, and equitable."[28]

Conference presenters spoke of the power of big data to tackle an array of social issues, from gender equity to climate resilience. Disaster recovery specialists explained how mobile finance and credit-card transaction data can help city leaders prevent price gouging after hurricanes and other extreme weather events, suggesting that mobile data could allow hurricane victims to find gas and groceries or assess who is creditworthy in a post-disaster setting. Catchphrases such as "When you have data that informs, you have data that transforms" or "The power of data is to drive good decisions based on fact and not politics" were frequently invoked to emphasize how private big data can catalyze social change.

Many of the #D4GX presentations described their initiatives in terms of "data philanthropy," an emerging practice whereby corporations donate data or insights generated from their data to the public (or a public-serving analyst such as a nonprofit institution) to yield new insights that could improve public policies or social programs and services. In addition to providing "evidence-based, data-driven" insights, data philanthropy intends to align business and philanthropic activities in a "shared value" strategy whereby companies link corporate social responsibility with competitive advantage to create social and economic value.[29]

The origins of data philanthropy can be traced to the 2009 World Economic Forum annual meeting in Davos where, in the aftermath of the global financial crisis, executives, government officials, and development experts introduced the idea of big data as an untapped resource for human well-being. In a series of reports following the Davos meeting, the WEF and UN Global Pulse introduced the principles of its project to build a new "ecosystem" of personal data management.[30] The new ecosystem was designed to respond to three main concerns: (a) creating value, (b) managing risk, and (c) strengthening trust. We describe below how these concerns were expressed.

Table 7.2. Data for Climate Action Documentation

Author	Year	Document Name	Document Type	Source
BBVA	2017	Data and Cybersecurity, Keys to Generate Confidence in a Digital Bank	Blog/News	https://www.bbva.com/en/data-cybersecurity-keys-generate-confidence-digital-bank/
BBVA	2017	BBVA to Join UN Global Pulse in "Data for Climate Action" Challenge	Blog/News	https://www.bbva.com/en/bbva-join-un-global-pulse-data-climate-action-challenge/
BBVA	2019	Big Data Contributes to the Sustainable Development Goals	Blog/News	https://www.bbva.com/en/big-data-contributes-to-the-sustainable-development-goals/
Bloomberg	2019	Data for Good Exchange 2019 Preview: Planet Track	Blog/News	https://www.techatbloomberg.com/blog/data-for-good-exchange-2019-preview-planet-track/
Business Wire	2013	Crimson Hexagon Partners with Public Sector Organizations through Social Research Grant Program	Blog/News	https://www.businesswire.com/news/home/20131025005507/en/Crimson-Hexagon-Partners-Public-Sector-Organizations-Social
Business Wire	2017	UN Global Pulse and Western Digital Host Data for Climate Event at 8th Sustainable Innovation Forum During COP23	Blog/News	https://www.businesswire.com/news/home/20171108005602/en/
Center for Strategic & International Studies	2017	Harnessing the Data Revolution to Achieve the Sustainable Development Goals	Report	https://www.csis.org/analysis/harnessing-data-revolution-achieve-sustainable-development-goals
CNN	2011	Crunching Digital Data Can Help the World	Opinion	http://edition.cnn.com/2011/OPINION/02/02/wolfe.gunasekara.bogue.data/index.html?section=cnn_latest
Data-Pop Alliance	2015	Big Data for Climate Change and Disaster Resilience: Realising the Benefits for Developing Countries	White Paper	https://datapopalliance.org/item/dfid-big-data-for-resilience-synthesis-report/

Source	Year	Title	Type	URL
Data-Pop Alliance	2016	Using Big Data to Detect and Predict Natural Hazards Better and Faster	Blog/News	https://datapopalliance.org/using-big-data-to-detect-and-predict-natural-hazards-better-and-faster-lessons-learned-with-hurricanes-earthquakes-floods/
Earth Imagining Journal	2015	Planet Labs Commits $60 Million in Geospatial Imagery to Global Community	News	https://eijournal.com/news/business-2/planet-labs-commits-60-million-in-geospatial-imagery-to-global-community
Fast Company	2011	Data Philanthropy: Open Data for World-Changing Solutions	Blog/News	https://www.fastcompany.com/1678963/data-philanthropy-open-data-for-world-changing-solutions
Fast Company	2017	How Mastercard's "Data Philanthropy" Program Is Tackling the Global Financial Information Gap	Blog/News	https://www.fastcompany.com/4045790/how-mastercards-data-philanthropy-program-is-tackling-the-global-financial-information-gap
Forbes	2011	Data Philanthropy Is Good for Business	Blog/News	https://www.forbes.com/sites/oreillymedia/2011/09/20/data-philanthropy-is-good-for-business/#32da89465870
Forbes	2015	Data Collaboratives: Sharing Public Data in Private Hands for Social Good	News	https://www.forbes.com/sites/bethsimonenoveck/2015/09/24/private-data-sharing-for-public-good/#1bdf1afe51cd
Forbes	2017	Mastercard's Big Data for Good Initiative: Data Philanthropy on the Front Lines	Blog/News	https://www.forbes.com/sites/ciocentral/2017/08/07/mastercards-big-data-for-good-initiative-data-philanthropy-on-the-front-lines/#3310288b20dc
Forbes	2019	Bloomberg's Data Initiatives: Big Data for Social Good in 2018	Blog/News	https://www.forbes.com/sites/ciocentral/2018/01/02/bloombergs-data-initiative-big-data-for-social-good-in-2018/#17bd48423a44

(continued)

Table 7.2. *Continued*

Author	Year	Document Name	Document Type	Source
GIZ	2017	Data for Development: What's Next? Concepts, Trends and Recommendations for German Development Cooperation	Report	http://webfoundation.org/docs/2018/01/Final_Data-for-development_Whats-next_Studie_EN.pdf
Global Partnership for Sustainable Development Data	2017	A World-Changing Combination: Dr. Claire Melamed on Big Data, Collaboration and the SDGs	Interview	http://www.data4sdgs.org/news/world-changing-combination-dr-claire-melamed-big-data-collaboration-and-sdgs
Global Partnership for Sustainable Development Data	2018	Climate Change Open Data for Sustainable Development: Case Studies from Tanzania and Sierra Leone	Report	http://www.data4sdgs.org/sites/default/files/services_files/WRI%20Climate%20Data_FINAL2_optimized.pdf
Global Partnership for Sustainable Development Data	2019	Partners Survey Results	Report	http://www.data4sdgs.org/sites/default/files/services_files/Partners%20Survey%20Report%202019.pdf
Global Partnership for Sustainable Development Data	2019	Could a Digital Ecosystem for the Environment Have the Potential to Save the Planet?	Blog/News	http://www.data4sdgs.org/news/could-digital-ecosystem-environment-have-potential-save-planet
GlobeNewswire	2014	UN Global Pulse & DataSift Annmounce Global Partnership	Blog/News	https://www.globenewswire.com/news-release/2014/07/02/1219319/0/en/UN-Global-Pulse-DataSift-Announce-Data-Philanthropy-Partnership.html

Source	Year	Title	Type	URL
Harvard Business Review	2013	A New Type of Philanthropy: Donating Data	Blog/News	https://hbr.org/2013/03/a-new-type-of-philanthropy-don
Harvard Business Review	2014	Sharing Data Is a Form of Corporate Philanthropy	Blog/News	https://hbr.org/2014/07/sharing-data-is-a-form-of-corporate-philanthropy
International Institute for Sustainable Development	2018	Big Data for Resilience	Report	https://www.iisd.org/publications/big-data-resilience-storybook
IT for Change	2019	Platform Planet: Development in the Intelligence Economy	Report	https://itforchange.net/report-platform-planet-development-intelligence-economy
Kindornay, S., Bhattacharya, D., and Higgins, K.	2016	Implementing Agenda 2030: Unpacking the Data Revolution at Country Level	Report	http://www.data4sdgs.org/sites/default/files/2017-09/Implementing%20Agenda%202030%20-%20Unpacking%20the%20Data%20Revolution%20at%20Country%20Level.pdf
Mastercard	2016	Donation Insights Helps Address the Information Gap Facing Charitable Organizations	Press Release	https://newsroom.mastercard.com/press-releases/donation-insights-helps-address-the-information-gap-facing-charitable-organizations/
Mastercard Center for Inclusive Growth	2016	A Call to Action on Data Philanthropy	Blog/News	https://www.linkedin.com/pulse/call-action-data-philanthropy-shamina-singh
Mastercard Center for Inclusive Growth	2018	Data for Good: Bringing Retail to Chicago's South Side	Blog/News	https://www.mastercardcenter.org/insights/data-snapshot-bringing-retail-chicagos-south-side

(continued)

Table 7.2. *Continued*

Author	Year	Document Name	Document Type	Source
Mastercard Center for Inclusive Growth	2018	Data Philanthropy Offers New Avenues for Solving Old Problems	Blog/News	https://www.mastercardcenter.org/insights/data-philanthropy-offers-new-avenues-solving-old-problems-report-finds
Medium	2017	John Snow Labs Expands Data Philanthropy Program and Joins the 1% Pledge Corporate Philanthropy Movement	Blog/News	https://medium.com/@JohnSnowLabs/john-snow-labs-expands-data-philanthropy-program-and-joins-the-1-pledge-corporate-philanthropy-235ec5ac9429
Open Data Watch	2018	Understanding the Impact and Value of Data	Blog/News	https://opendatawatch.com/blog/understanding-the-value-impact-of-data/
Orange	2018	Data Sharing, an Environmental Issue	Interview	https://www.orange.com/en/newsroom/news/2020/data-sharing-environmental-issue
Philanthropy Daily	2014	The Latest Call for Data Philanthropy	Opinion	https://www.philanthropydaily.com/the-latest-call-for-data-philanthropy/
PR Newswire	2013	Teradata Announces New Focus on Data Philanthropy	Blog/News	https://www.prnewswire.com/news-releases/teradata-announces-new-focus-on-data-philanthropy-223903271.html
PR Newswire	2013	Teradata Honored by the White House for Leadership	Blog/News	https://www.prnewswire.com/news-releases/teradata-honored-by-the-white-house-for-leadership-in-bringing-big-data-analytics-to-governments-and-non-profits-231587871.html
PR Newswire	2017	Mastercard Receives 2017 Global Shared Value Award	News	https://www.prnewswire.com/news-releases/mastercard-receives-2017-global-shared-value-award-300551061.html
PSFK	2012	Future of Real-Time Information	Report	https://www.slideshare.net/PSFK/psfk-presents-future-of-realtime

Pulse Lab Jakarta	2014	International Conference. Data Innovation for Policy Makers	Conference Proceedings	http://unglobalpulse.org/sites/default/files/ Proceedings%20Data%20Innovation%20 Conference.pdf
Pulse Lab Jakarta	2017	Data Revolution for Policy Makers. International Conference	Report	https://issuu.com/pulselabjakarta/docs/drfp_2017_ proceeding
Pulse Lab Kampala	2016	Catalyzing a Responsible "Big Data for Development Ecosystem"	Report	https://www.unglobalpulse.org/wp-content/ uploads/2016/05/PLK-Track-2-progress-report-2015-2016.pdf
Pulse Lab Kampala	2017	Pulse Lab Kampala Progress Report 2016–17	Report	https://www.unglobalpulse.org/sites/default/files/ PLK-FINAL-ANNUAL-PROGRESS-2017-220118. pdf
SciDevNet	2013	UN Initiative Mines Big Data to Direct Development	Blog/News	https://www.scidev.net/global/data/news/un-initiative-mines-big-data-to-direct-development. html
SciDevNet	2014	Big Data for Development: Facts and Figures	Blog/News	https://www.scidev.net/global/data/feature/big-data-for-development-facts-and-figures.html
The Chronicle of Philanthropy	2017	A MasterCard Plan for Financial Inclusion	Podcast	https://philanthropynewyork.org/news/ podcast-mastercard-plan-financial-inclusion
The Guardian	2015	Aimia Harnesses the Power of Data Insight	Opinion	https://www.theguardian.com/sustainable-business/ 2015/apr/30/aimia-harnesses-the-power-of-data-insight-for-social-good
The Times (London)	2017	Data Is the Real Cryptocurrency Big Corporations Cannot Get Enough Of	News	https://www.thetimes.co.uk/article/data-is-the-real-cryptocurrency-big-corporations-cannot-get-enough-of-0ml3tsb0n

(continued)

Table 7.2. *Continued*

Author	Year	Document Name	Document Type	Source
UN ECOSOC Partnership Forum	2018	Partnering for Resilient and Inclusive Societies: Contributions of the Private Sector	Report	https://www.un.org/ecosoc/sites/www.un.org.ecosoc/files/files/en/2018doc/2018-partnership-forum-summary.pdf
UN ESCAP	2015	Big Data and the 2030 Agenda for Sustainable Development	Report	https://www.unescap.org/sites/default/files/Final%20Draft_%20stock-taking%20report_For%20Comment_301115.pdf
UN Global Pulse	2011	Rapid Impact and Vulnerability Analysis Fund (RIVAF) Final Report	Report	https://www.unglobalpulse.org/document/rapid-impact-and-vulnerability-analysis-fund-final-report/
UN Global Pulse	2011	Data Philanthropy: Public & Private Sector Data Sharing for Global Resilience	Blog/News	https://www.unglobalpulse.org/blog/data-philanthropy-public-private-sector-data-sharing-global-resilience
UN Global Pulse	2012	Big Data and Real-Time Analytics for Agile Global Development	Report	https://beta.unglobalpulse.org/document/big-data-and-real-time-analytics-for-agile-global-development/
UN Global Pulse	2012	Big Data for Development: Challenges & Opportunities	Report	http://www.unglobalpulse.org/sites/default/files/BigDataforDevelopment-UNGlobalPulseMay2012.pdf
UN Global Pulse	2012	Taking the Global Pulse—Using New Data to Understand Emerging Vulnerability in Real-Time	Report	https://beta.unglobalpulse.org/wp-content/uploads/2012/02/GlobalPulseBook-FINALPRINT.pdf
UN Global Pulse	2013	Big Data for Development: A Primer	Report	http://www.unglobalpulse.org/sites/default/files/Primer%202013_FINAL%20FOR%20PRINT.pdf
UN Global Pulse	2013	Mobile Phone Network Data for Development	Report	http://www.unglobalpulse.org/sites/default/files/Mobile%20Data%20for%20Development%20Primer_Oct2013.pdf

UN Global Pulse	2013	Data Philanthropy: Where Are We Now?	Blog/News	https://www.unglobalpulse.org/data-philanthropy-where-are-we-now
UN Global Pulse	2014	Annual Report 2013	Report	https://www.unglobalpulse.org/2014/02/global-pulse-annual-report-2013/
UN Global Pulse	2015	Annual Report 2014	Report	https://www.unglobalpulse.org/2015/04/global-pulse-annual-report-2014/
UN Global Pulse	2015	Improving Data Privacy & Security in ICT4D	Report	https://www.unglobalpulse.org/document/improving-data-privacy-data-security-in-ict4d-meeting-report/
UN Global Pulse	2015	Big Data for Development in Action: Global Pulse Project Series	Report	https://www.unglobalpulse.org/document/big-data-for-development-in-action-un-global-pulse-project-series/
UN Global Pulse	2016	Annual Report 2015	Report	https://reliefweb.int/report/world/global-pulse-annual-report-2015
UN Global Pulse	2016	The Importance of Big Data Partnerships for Sustainable Development	Blog/News	https://www.unglobalpulse.org/big-data-partnerships-for-sustainable-development
UN Global Pulse	2017	Annual Report 2016	Report	https://www.unglobalpulse.org/wp-content/uploads/2016/05/UNGP-Report-2016_DIGITAL-VERSION.pdf
UN Global Pulse	2017	The State of Mobile Data for Social Good	Report	http://unglobalpulse.org/sites/default/files/MobileDataforSocialGoodReport_29June.pdf
UN Global Pulse	2017	These Are the Winners of the Data for Climate Action Challenge	Blog/News	https://www.unglobalpulse.org/2017/11/these-are-the-winners-of-the-data-for-climate-action-challenge/

(continued)

Table 7.2. *Continued*

Author	Year	Document Name	Document Type	Source
UN Global Pulse	2018	Annual Report 2017	Report	https://www.unglobalpulse.org/document/un-global-pulse-2017-annual-report/
UN Global Pulse	2018	Experimenting with Big Data and Artificial Intelligence to Support Peace and Security	Report	https://www.slideshare.net/unglobalpulse/experimenting-with-big-data-and-ai-to-support-peace-and-security
UN Global Pulse	2019	Annual Report 2018	Report	https://www.unglobalpulse.org/sites/default/files/UNGP_Annual2018_web_FINAL.pdf
UN Global Pulse	2019	Unpacking the Issue of Missed Use and Misuse of Data	Blog/News	https://www.unglobalpulse.org/2019/03/unpacking-the-issue-of-missed-use-and-misuse-of-data/
UN Global Pulse	2020	Data for Climate Action	Blog/News	https://www.unglobalpulse.org/data-for-climate-action
UN News	2015	The UN Body with Its Finger on the Pulse of Sustainable Development	News	https://news.un.org/en/story/2015/07/503262-feature-un-body-its-finger-pulse-sustainable-development
UN News	2019	UN Makes 'Declaration of Digital Interdependence', with Release of Tech Report	News	https://news.un.org/en/story/2019/06/1040131
UN Secretary-General's High-Level Panel on Digital Cooperation	2019	The Age of Digital Interdependence: Report of the High-Level Panel on Digital Cooperation	Report	https://www.un.org/en/pdfs/DigitalCooperation-report-for%20web.pdf

	Year	Title	Type	URL
UN Secretary-General's Independent Expert Advisory Group on a Data Revolution for Sustainable Development (IEAG)	2014	A World That Counts: Mobilising the Data Revolution for Sustainable Development	Report	https://www.undatarevolution.org/wp-content/uploads/2014/11/A-World-That-Counts.pdf
IEAG	2014	You Say You Want a Revolution	Blog/News	https://www.undatarevolution.org/2014/10/08/say-want-data-revolution/
IEAG	2014	The Data Revolution for Human Development	Blog/News	https://www.undatarevolution.org/2014/11/07/data-revolution-human-development/
UN Sustainable Development Goals	2015	UN Bodies Present Project Showing How "Big Data" Can Save Lives, Fight Hunger	Blog/News	https://www.un.org/sustainabledevelopment/blog/2015/04/un-bodies-present-projects-showing-how-big-data-can-save-lives-fight-hunger/
UN Sustainable Development Goals	2016	Twitter, UN Global Pulse Announce Data Partnership	Blog/News	https://www.un.org/sustainabledevelopment/blog/2016/09/twitter-and-un-global-pulse-announce-data-partnership/
United Nations	2017	"Connect the Unconnected," Deputy Secretary-General Tells Digital Technology Panel, Urging Full Inclusion to Advance Societies Everywhere	Press Release	https://www.un.org/press/en/2017/dsgsm1049.doc.htm
Urban Institute	2018	Data Philanthropy: Unlocking the Power of Private Data for Public Good	Report	https://www.urban.org/sites/default/files/publication/98810/data_philanthropy_unlocking_the_power_of_private_data_for_public_good_2.pdf

(continued)

Table 7.2. *Continued*

Author	Year	Document Name	Document Type	Source
Western Digital	2017	UN Global Pulse and Western Digital Announce "Data for Climate Action" Challenge Now Open for Entries	Blog/News	https://www.westerndigital.com/company/newsroom/press-releases/2017/2017-03-09-un-global-pulse-and-western-digital-announce-data-for-climate-action
Western Digital	2017	UN Global Pulse and Western Digital Announce Winners of "Data for Climate Action" Challenge	Blog/News	https://www.westerndigital.com/company/newsroom/press-releases/2017/2017-11-29-un-global-pulse-and-western-digital-announce-winners-of-data-for-climate-action-challenge
Western Digital	2018	Data-Directed Road Repairs Could Save Money and Lives	Blog/News	https://datamakespossible.westerndigital.com/data-directed-road-repairs-save-money-lives/
Western Digital	2017	What Is Data Philanthropy?	Blog/News	https://datamakespossible.westerndigital.com/what-is-data-philanthropy/
Western Digital	2017	Data Innovation: Generating Climate Solutions	Video	https://datamakespossible.westerndigital.com/data-innovation-generating-climate-solutions/
World Bank	2018	Information and Communications for Development 2018: Data-Driven Development	Book	https://openknowledge.worldbank.org/handle/10986/30437
World Economic Forum (WEF)	2011	Personal Data: The Emergence of a New Asset Class	Report	http://www3.weforum.org/docs/WEF_ITTC_PersonalDataNewAsset_Report_2011.pdf
WEF	2012	Big Data, Big Impact: New Possibilities for International Development	Report	http://www3.weforum.org/docs/WEF_TC_MFS_BigDataBigImpact_Briefing_2012.pdf
WEF	2014	Rethinking Personal Data: A New Lens for Strengthening Trust	Report	https://reports.weforum.org/rethinking-personal-data/
WEF	2015	Paving the Path to a Big Data Commons	Report	http://vitalwave.com/wp-content/uploads/2015/09/Paving-Path-Big-Data.pdf

(a) *Creating Value.* A central objective of the WEF and its partners was to promote personal data as a valuable economic resource in a post-industrial (and post-financial crisis) environment. Echoing the "data is oil" metaphor, WEF and UN Global Pulse reports highlighted the potential for innovation, real-time connectivity, and "unprecedented" global reach of big data insights. Key to the achievement of this value was a multi-stakeholder approach, in which all parties to the transaction (as well as all of the datasets each party could contribute) could be mobilized in the service of collective gains. A personal data "ecosystem" was therefore imagined as a way to " 'balance' the needs of government, private industry and individuals in order to create value."[31] In the case of data for sustainable development, a data ecosystem that brought together the "disparate worlds of public, private and civil society data" to "develop a global consensus on principles and standards" was especially important to promote big data as a source of inclusion and equality.[32]

(b) *Managing Risk.* Despite, or perhaps because of, the WEF's elaborate claims to economic and social value, the organization was well aware of the need to account for the mounting anxieties of users and national governments over the privacy and security of their data. The greatest concern for the WEF and its partners was to maintain the "opportunity" structure of personal data collection and targeting while minimizing the risks (or at least the appearance of risk) to users and regulators. The reports therefore proposed a perspective that *distributed* risk among the various stakeholders, arguing that the "balance" created by a multi-stakeholder ecosystem model would overcome uncertainties. Three kinds of risk were identified: "the risks of private sector imbalance," by which companies become overcompetitive in their quest for user data and decrease user trust; "the risk of public sector imbalance," by which national governments "inadvertently stifle value creation by overregulating" data collection and surveillance, "slowing down innovation and investment"; and "the risk of end user imbalance," by which individuals, "in the absence of engagement with both governments and business . . . self-organize and create non-commercial alternatives for how their personal data is used."[33]

(c) *Strengthening Trust.* In response to these risks, and in the shadow of increasing political debates over the need to regulate technology companies' data collection, by May 2014, the WEF and its collaborators had seeded the establishment of "trust networks and holistic incentive structures" among development agencies and the private sector "to facilitate data exchange but also to ensure that risk management is held to the highest standard."[34]

In a 2014 World Economic Forum report, "Rethinking Personal Data: A New Lens for Strengthening Trust," the authors explained how these trust networks would be maintained by an approach to transparency, accountability, and empowerment that was not universal or omnipresent but rather situated and contextual. Arguing that a contingent relationship to such values constituted an evidence-based approach, the report's authors emphasized the unique properties of data-derived information to provide "real" insights for environmental governance—filling in the holes the "crafty science" climate researchers use to evaluate incomplete datasets and simulations.[35]

For the protagonists of this data ecosystem, transparency had to be made *meaningful* in order to accrue value. "Meaningful transparency," as the WEF report called it, was tied up with a strategic approach to publicity.[36] In some contexts, private sector actors need to appear transparent or accountable in a given situation to gain public trust; but if "evidence" of transparency or accountability is not required, it is wiser to maintain a distance from public scrutiny. Just as James Grunig's model of situational publics presented a world in which problems only became problems when publics were incentivized to care about them, the trust networks imagined by the World Economic Forum and its peers operated along an incentive structure that appealed to self-interest as the motivation for attention and consideration.

These trust networks, with WEF and UN Global Pulse at their core, would find additional partners to help them frame ongoing data collection as a public good, while maintaining the promise of benefits to all parties and generating economic value for private sector participants:

> Aligning the different interests to create a true "win-win-win" state for all stakeholders presents a challenge—but it can be done. The solution lies in developing policies, incentives and rewards that motivate all stakeholders— private firms, policy makers, end users—to participate in the creation, protection, sharing and value generation from personal data.[37]

In the next section, we draw on findings from our interviews and participant observation to show how these principles of the data ecosystem have played out in the promotion of data initiatives for climate change action.

UN Global Pulse and the
Data for Climate Action Campaign

The United Nations foresaw the rise of big data analytics as an opportunity to support the achievement of its Sustainable Development Goals (SDGs). In 2013, UN Secretary-General Ban Ki-moon authorized the formation of an Independent Expert Advisory Group on a Data Revolution for Sustainable Development. In November 2014, the Advisory Group released its first report, "A World That Counts: Mobilizing the Data Revolution for Sustainable Development." The report makes three cases for big data as a crucial support to the achievement of the SDGs. First, it positions big data as an appropriate technological intervention that can "paint a richer picture of human development," one where, for instance, a measurement like the Human Development Index could be expanded to include alternative development dimensions like "voice, equality, sustainability, freedom and dignity."[38] Second, big data is presented as a complement to national statistical systems, increasing the diversity and accessibility of relevant data that can lead to better dialogues and decision-making. Third, the report proposes that big data could "move the world onto a path of information equality," where every government, organization, and citizen can access—and be accountable to—the knowledge it generates.[39]

To unlock the capacity of big data and data analytics to provide insights into sustainable development problems, gaining the participation of the private sector was key. This is the role of the United Nations Global Pulse, an "innovation lab" created in 2009 to "bring . . . together expertise from inside and outside the UN to harness today's new world of digital data and real-time analytics for global development."[40] A vocal proponent of the "data for good" model, Global Pulse, for the last ten years, has sought out private sector data partnerships with companies such as social media businesses and mobile telecommunications operators.[41] Global Pulse's vision has been to build "a future in which big data is harnessed safely and responsibly as a public good" through the promotion of data philanthropy and other kinds of private-public collaboration. A Global Pulse director described the organization's raison d'être:

> Global Pulse is the result of the only request the G20 ever made to the UN— it's not a well-known fact but it's interesting. . . . Most of what we do is not really about measuring progress. This is not about generating statistical indicators. It's about smarter implementation of programs and more effective management of risk. This is really about looking at how we can use digital evidence of human behavior to make reliable inferences about what's happening offline at the household level. (Respondent 7)

This description highlights the private sector orientation of the organization as well as its mission to "innovate" by generating alternative means of approach to public good problems.

Creating Value: Shifting Regimes of Expertise

Since a central mission of UN Global Pulse was to tout the unique expertise of the private sector in the resolution of public problems, the agency tried to downplay its own authority as an intergovernmental organization. To elevate the perceived value of company data and the unique expertise of private data owners, UN Global Pulse repositioned itself as a sort of network facilitator— a "safe partner" where companies could "work in a sandbox" and explore the applicability of corporate data to advancing the SDGs. In its role as a partnership broker between UN agencies and data companies, UN Global Pulse advises them on how to navigate the institutional, legal, and economic barriers to using privately owned big data for the public good. Through trial and error, Global Pulse has been able to refine the concept of data philanthropy and promote it as a valuable public-private partnership in the data economy. A director at UN Global Pulse characterized the organization's approach this way:

> In the early days, none of them knew how to do any of this, so we were like— not that we were making it up as we went [along], but—in other words, UNICEF would say, "Well, we're interested in doing this project." We'd be like, okay. Let's go out and partner with Twitter. And let's make sure we have someone on staff who knows how to do sentiment analysis, and how do we

coordinate with UNICEF to make sure we get the expertise in knowing what to look for? We're basically doing full-cycle—it's joint concept development, but like full cycle project management, and every aspect of it was us. After a few of those, UNICEF's like, this is cool. We get it. Can you help us get a conversation going with a mobile operator in Tanzania? And then a year or two later it's like, can you help us hire a data scientist? And now they don't need us for anything. They're off and running, and that's the point. (Respondent 7)

To attract further partners and showcase the benefits of data for good to the research community and the general public, UN Global Pulse hosted two data "challenges," the "Big Data Climate Challenge" in 2014 and the "Data for Climate Action" challenge in 2017. In collaboration with the philanthropic foundation Skoll Global Threats Fund and Western Digital (an American data technology company), Global Pulse collected a number of datasets —"donated" by companies such as BBVA Data & Analytics, Orange Telecommunications, and Waze—and provided them to teams of researchers and data scientists who had volunteered their time to compete to identify opportunities contained in the data in the service of climate action (Sustainable Development Goal #13). The challenges were promoted on YouTube and in other media, effectively functioning as public relations for the notion of D4CA. Indeed, as a climate policy advisor with the UN Secretary-General's Climate Change Support Team explained,

We [the Support Team] had developed a strategy early on, that we wanted to flip the climate crisis on its head after Copenhagen [the 2009 UN Climate Change Conference] and reframe it as an opportunity for solutions, because the global community was getting apathetic. There was an idea that only governments could solve the problem and they had failed to do so in Copenhagen. And so we were trying to restructure the paradigm so that— you might now hear the phrase, "all hands on deck"—this is a crisis which is also an opportunity for everyone to be engaged at all levels to deliver solutions . . . so yeah, that's how [the D4CA challenges] came about, really trying to tap into a new community of actors and a new way of delivering solutions for the climate stakes. (Respondent 6)

The phrase, "all hands on deck," and the notion of crisis-as-opportunity underlie the strategy by which UN Global Pulse and its affiliates brought private sector companies on board to address climate change. By promoting climate change as a major opportunity for businesses to intervene, there needed to be an enforcement of the idea that business expertise specifically was urgently required. This view is echoed by "ecological modernization" advocates such as Maarten Hajer and his collaborators, who argue that multi-lateral environmental agreements have so far failed to meet their goals because of ongoing "cockpit-ism": a "top-down logic of steering" by which national leaders issue international policy directives from a "cockpit," limiting the authority of other actors to participate in decision-making.[42] Hajer and his co-authors strongly advocate the inclusion of business in decision-making around environmental policy, arguing that the "universal relevance" of climate change requires multiple participants in order to reach consensus around international action. To bring business on board, innovation and marketizability are important motives:

> The SDGs need to connect to the logic of the business and finance community, and mobilize and engage them as agents of change. This requires toning down the narrative of limits and emphasizing the narrative of opportunities.[43]

By focusing on business as "agents of change," UN Global Pulse and its affiliates could make the SDGs into "an influential and transformative norm in the 21st century."[44] The D4CA was one publicity element to make this happen.

At the same time, the business participants were clear about the stakes of their participation. While relatively convinced of the "data for good" model to which their participation adhered, the notion of "philanthropy" was not entirely accepted. Shared value, for data company participants in D4CA, may mean that other stakeholders benefit from their data, but not without a profit-generating motive. As the director of customer success at a participating data company explained:

> NGOs should have access to some of the same tools that corporations and business have. Their use cases sometimes are not that much different. I mean, if you're doing something on, let's say, the UN and climate change, you want to know for example [in] which countries do people think, "it's a hoax," and [in] which countries do people think some action can be taken. So you're trying to assess your audience and what they think about

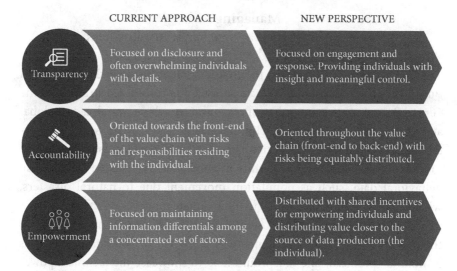

CURRENT APPROACH   NEW PERSPECTIVE

Transparency — Focused on disclosure and often overwhelming individuals with details. — Focused on engagement and response. Providing individuals with insight and meaningful control.

Accountability — Oriented towards the front-end of the value chain with risks and responsibilities residing with the individual. — Oriented throughout the value chain (front-end to back-end) with risks being equitably distributed.

Empowerment — Focused on maintaining information differentials among a concentrated set of actors. — Distributed with shared incentives for empowering individuals and distributing value closer to the source of data production (the individual).

Figure 7.1. "A New Lens for Strengthening Trust." *Source*: World Economic Forum, *Rethinking Personal Data*, 2014, 4.

something. It's not that much different than a company trying to sell a product, trying to understand which country would be more likely to buy that product, versus [in] which country would that not sell. So NGOs should, in a perfect world, have access to the same sophisticated tools that business has. The thing is that, as *we* are a business, we can't just give it away because there's a lot of support and time it takes to help a customer and if they're not paying, then we're just losing money on it and then we're not a viable business and we can't help anybody. (Respondent 16)

What became clear over the course of these interviews was that "data philanthropy" was a promotional, public-facing strategy. It was good PR to call data companies data "donors"; but the practical limitations companies expressed prevented their proprietary data from being freely distributed. A member of the D4CA evaluation committee articulated the problem: "In terms of the marketing languages, you needed a hook that was very compelling. . . . [Data philanthropy] was a great term, and you didn't for broad, awareness reasons want to pour cold water on that—but it was a term that was fraught" (Respondent 2). It just wasn't sustainable, the committee member explained, to give it away.

Managing Risk:
Climate Change as a "Safe Space" for Business

The D4CA 2014 challenge was the first data challenge launched by Global Pulse. The challenge theme, "climate action," was considered strategic by the Global Pulse team for a number of reasons. First, there was already a global audience interested in tackling climate change issues. There was also an established community of data scientists using public data to estimate climate models. Several of these community members were interested in exploring behavioral data, such as population movement due to natural disasters, which could eventually be added to their climate models.

Third, and perhaps most closely aligned to the image concerns within the business sector, climate change was presented to potential partners as a "neutral" problem, one that would help to showcase the technical power of big data without having to confront the political contention arising from data applications in conflict settings. A member of the D4CA Evaluation Committee based at the WEF put it this way:

A. [In the environment space] in general, the concerns are more macro, and it doesn't necessarily entail, if you will, instrumenting the social structures. So when you look at water rights, or farming, you're ultimately telling a group of people or businesses, "Okay, here's how you ought to do things differently." Versus, if it's more, you know, oceans data, or climate-related elements, you can kind of abstract a layer and say "Okay, here's how the *earth* [laughs] is changing."
Q. You take away the agency.
A. Right, and you're, to a degree, kind of smudging away the social tensions. (Respondent 2)

If one concern from within the business community was to avoid the potential "political" implications of private-sector data use and avoid the appearance of data imperialism, a second perceived risk was the privacy element. Here the UN Global Pulse, the World Economic Forum, and other movers in the D4CA challenge aimed to present climate change as an *unobtrusive* context for action by leaning on the idea of climate change as less "risky" for businesses because it would seem less as if user privacy was at stake:

Just grabbing an example . . . me on the Weather Channel app where I'm clicking and checking stuff like that. If some of that data could be fed into a UN OCHA [Office for the Coordination of Humanitarian Affairs] dataset that would only be used in the context of a natural disaster. Then, okay, I wouldn't feel that was an intrusion, right? (Respondent 2)

To transform climate change action from risk to opportunity, UN Global Pulse and the WEF hit on the idea of "missed use": the risk to businesses of *not* participating in decision-making around global climate change responses. The idea of missed use was formulated as a direct response to the GDPR. While the GDPR emphasized the need to protect users from the *misuse* of their data, UN Global Pulse emphasized the need for business to avoid missing out on the uses to which their data could be put. As Robert Kirkpatrick, the director of UN Global Pulse, wrote on the organization's website:

Just because data misuse is at the forefront of recent conversations, we shouldn't ignore the harms associated with missed use. Lost opportunities to use big data to achieve the Sustainable Development Goals are probably to blame for at least as much harm as leaks and privacy breaches.[45]

The data philanthropy framework was instrumental in presenting data for climate action in terms of "missed use." Leaning on the shared value paradigm, UN Global Pulse emphasized the social and economic benefit to companies of donating data to environmental causes, but especially the risks they faced if they did *not* take advantage of this potential. A UN Global Pulse senior staff member described at length how this idea was conveyed during a meeting with a technology platform company:

We had this conversation with PayPal a few years ago. They were like, "Our CSR priorities are around disaster resilience and financial inclusion." I'm like, yeah, totally, because your products—financial service products—directly fit into those two sectors in terms of outcomes, but guess what? We could use your data to understand how effective climate action is. And they were like, "We don't do climate." But now you [PayPal] have an ethical obligation to figure out whether you *should*. Because what if your greatest asset is actually seeing what people buy and sell in ways that interact with the climate system? And they [PayPal] were like, "Whoa." So there's that aspect of it. Then there's the companies that—you know, like a mobile operator who spends $3 billion building tower infrastructure in an emerging market and then there's a huge drought, and people are affected in ways that cause them to have to sell their assets and unsubscribe from the sports scores and weather updates and everything else you were counting on to monetize that infrastructure. There goes your business and—oh, but it turned out that in your data warehouse three months earlier, there were changing patterns of mobile consumption on a population movement that could have been used to identify those most vulnerable for cash transfers, for school feeding programs, for risk communication, for disaster preparedness. You just created business risk by *not* figuring out how to inform the policies that strengthen the economic resilience of your markets. (Respondent 7)

Trust in Numbers: Evidence-Based Decision-Making

The D4CA challenge invoked an evidence-based rhetoric that sees population data (or the lack thereof) as a means to justify action (or inaction) and policy interventions.[46] D4CA frames big data as the missing link in the policy-research chain;[47] a tool with unlimited possibilities that can surpass the limitations of traditional survey methods and fill gaps in regions where the lack of accurate and timely data delays the achievement of SDGs:

Surveys are a high-resolution picture. They provide very good resolution . . . but it's a picture. It's a snapshot of what happened in a specific area at a time. Big data analytics, however, are more like a webcam—they're moving scenery of what's going on in real time. Not necessarily the best in terms of resolution but enough to give you an idea of what's going on and enough to give you an idea of whether something's going very wrong or not. (Respondent 9)

The D4CA challenges positioned data philanthropy as a convenient and inexpensive means to access data that are not publicly available, and as a way to complement and eventually improve national statistical systems. As we have seen, the private sector outreach was a major factor for D4CA as well. A member of the D4CA Technical Committee said:

There's lots of stuff we don't know and lots of information that's impossible to collect or too expensive to collect. Or you can't go into this area because it's a conflict area or you can't find these people because they're marginalized. So there's this idea that big data could fill the gaps in our information based off official statistics, coupled with a desire to work with the private sector, coupled with a desire to be on the cutting edge, coupled with—being a couple years after the emergence of social media and that kind of massive explosion of data, as well as some very prominent examples of corporations and the private sector using big data. So I think all those came together to drive a lot of attention for [the D4CA challenges]. (Respondent 12)

Some interviewees—especially those working in intergovernmental organizations and research centers—acknowledged risks in terms of data quality, interpretation, and representation. They recognized the possible distortions of an overreliance on numbers as a means to recognize, incorporate, or govern vulnerable populations and showed concern over the tensions between data companies' goals and those of national or regional governments.[48] One respondent engaged in a hypothetical to explain what might happen if data companies were held responsible for creating infrastructure to serve entire populations:

So, there's interesting population statistics, and you can then infer a variety of other policy questions. Some of those policy questions may have a direct commercial impact on some of the data holders. And, just arbitrarily making something up, [if the Senegalese government said to a data company] "Hey, in this part of the country nobody's getting access to your infrastructure. So go build infrastructure, because you're not serving every citizen of Senegal, and we don't really care about your commercial return. These people [are] being underserved because there's no data accessible to them." So then, maybe [data companies] don't want to share, because then, all of a sudden, you can see a bias. (Respondent 2)

Conclusion

In 2018, the UN Economic and Social Council (ECOSOC) Forum, which brings together over 300 representatives of member states and a wide range of non-state actors, considered the issue of private-public collaboration in the age of big data. ECOSOC concluded that big data is a valuable "business asset that the private sector can donate to governments for more informed public policy-making."[49] In 2019, the High-Level Panel on Digital Cooperation— established by UN Secretary-General António Guterres and chaired by philanthropist Melinda Gates and chairman of the Alibaba Group, Jack Ma— further explored the issue of private-sector data sharing and proposed more concrete alternatives for private-public cooperation in the data economy.[50] Chapter 4 of the High-Level Panel on Digital Cooperation report, titled "The Age of Digital Interdependence," outlines three potential mechanisms for digital cooperation: one that builds on the multi-stakeholder Internet Governance Forum; one that proposes a distributed architecture building on existing UN mechanisms, and a third that advocates a "data commons" approach with little coordination by the UN.[51]

This latter mechanism is at work in the report, "Sharing Is Caring: Four Key Requirements for Private Data Sharing and Use for Public Good," by the Data-Pop Alliance. The Data-Pop Alliance, a big data think tank made up of academic researchers, data scientists, and development experts, among others, was founded by a former UN Global Pulse member, Emanuel Letouzé. As he explained in an interview in 2015, Letouzé had become disenchanted with the UN Global Pulse efforts. "I didn't think the 'techno-scientific' approach and the 'data-for-good' narrative they embodied would make much of a difference. I thought it overlooked many aspects of the problems the world faces."[52]

The Data-Pop Alliance claims a more cautious stance, attempting to parse the difference between data sharing and ethical considerations. In situations of global health or humanitarian crisis, the goal is to save lives; and by this metric the idea that data *must* be shared is a powerful exhortation. But for some, the question is not how more information, delivered more quickly to more places, can be mobilized in an "all hands on deck," consensus-oriented approach; it is to consider which human rights are most fundamental to our humanity. There is, as Letouzé argues, a considerable tension between different rights—the right to privacy, for instance—and especially between what we "can" do and what we "should" do to resolve global public problems. Climate change is by its nature a crisis of global proportions. It demands widespread participation to recognize it as a problem and to devise collective responses to apprehend it:

> Our [Data-Pop Alliance's] stance is that in modern pluralistic data-infused societies, the most fundamental human right is political participation,

specifically the right and ability of citizens and data producers to weigh in on debates about what constitutes a harm, notably through greater legal and effective control over the rights and use of their data. This perspective highlights the fundamental political nature and requirements of the (Big) Data Revolution—one that is about people's empowerment, not just about the ability of politicians and corporations to get and use or misuse more individual data.[53]

Still, Data-Pop Alliance, like many of the other data-driven organizations and coalitions acting on environmental problems, struggle to account for the larger problem at hand: the super-wicked problem of climate change is not the domain of "stakeholders"; it is not a matter of "contextual" formulation; and it will not bend to individual "empowerment." Climate change requires not a situational approach but a transformed nature of being. This is not possible in a data-delimited commons, where sustainability is more likely to refer to the legacy of data-driven problem-solving than to the commitment to environmentally safe futures.

Environmental information systems structure what we see in the environment. They help determine what the problems are and what means we might draw on to solve them. They enable a certain kind of legitimacy, one that is used for policy determinations and practical approaches to resolving the problems as they have been designed. What we learn from campaigns like Data for Climate Action is that they design problems in the image of those who stand most to benefit from solving them. In their bid to render climate change more "meaningful" for interested stakeholders in realms of publicity, they operate more as promotional techniques to gain acceptance for their application than as necessary interventions in the global public crisis of climate change.

Conclusion

We're Supposed to Be Engaging

We're in a democracy, right? We're supposed to be engaging.
—Richard S. Levick

The room was mostly quiet at the Oil & Gas Public Relations and New Media Conference in National Harbor, Maryland, as Richard Levick delivered his keynote address. It was May 2015, eighteen months before the election of Donald Trump to the White House would set in motion a series of efforts to destroy federal environmental data and dismantle the environmental regulations, budgets, and research put in place over the preceding decades.[1] The conference attendees—campaign strategists, oil and gas company communications directors, political staffers, trade media, industry council groups, lobbyists, marketers, and PR professionals—were gathered around tablecloth-covered tables, wedding style, to hear Levick's presentation about the "reputational challenges and opportunities" of new media for the industry.

The conference site lay just south of the nation's capital along the shore of the Potomac River. The Potomac was a strategic waterway during the American Civil War, the war fought to preserve the democratic system of government enacted in the Constitution. Today, National Harbor is a massive mixed-use waterfront development, with a "planned community" of townhouses, "manor homes," and condominiums as well as 350 acres of resort space with shops, restaurants, a golf course, and a casino.[2] Controversy in the late 1990s over the environmental hazards of the development—"aquatic impact, environmental justice concerns, and air quality/transportation questions"—had been tamped down by a legislative rider that excluded the developers from having to complete an environmental impact statement.[3]

It's not clear if Levick was aware of the conference site's symbolic potential. But his speech suggested a deep concern with the idea of democracy and with the seeming transformation of the system of social and economic progress that had supported the conditions of modern public life. In the 1980s, after completing a master's degree in environmental advocacy at the University of Michigan, Levick worked for the Michigan Public Interest Research Group

A Strategic Nature. Melissa Aronczyk and Maria I. Espinoza, Oxford University Press. © Oxford University Press 2022.
DOI: 10.1093/oso/9780190055349.003.0009

(PIRG), one of dozens of state organizations set up by the environmental and consumer advocate Ralph Nader to monitor members of Congress, lobby in city halls and state legislatures, and prepare national campaigns to influence political leaders and hold them accountable to their claims. At the time, Nader's commitment to organizing citizens dedicated to finding "common ground" around public problems for a "healthier, safer world" left an impression on Levick. During a visit by Nader to the Michigan PIRG in 1982, Levick told the *Washington Post* that Nader was his hero, insisting, "He's the only one in Washington I'd like to be like."[4]

Four decades later, Levick is an established Washington public affairs specialist, litigation strategist, and crisis communications expert, with a long list of high-profile industry and government clients and an eponymous PR firm whose slogan is "When you need to make the problem go away." His background as one of "Nader's raiders" is a source of pride, frequently mentioned in his speeches and interviews. It has become a currency of legitimacy for his work with clients overwhelmed by the pressures of public opinion in environmental and other high-conflict arenas.[5]

At this PR event, it was clear Levick was a seasoned speaker, and he knew his audience. There were wan smiles and nods of approval and the occasional "ohhhs" of recognition as he weaved through the tables, pausing occasionally to jab a finger at an attendee as he built his persuasive case. The problem today, explained Levick, was that we—the "we" meant to take in the oil and gas industry in general, and those charged with promoting its social and economic benefits in particular—were doing a terrible job trying to engage with our publics.

> So we have this entire revolution that has taken place and we are not participating. Where are we on our great issues, the issues that we care about? Where are we on fracking? Where are we on Keystone [the Keystone XL pipeline]? On Keystone we have spent . . . we have 50 MP [midstream pipeline] companies spending $128 million that five environmental departments and 12 environmental groups have spent, opposing it— according to the Lobbying Disclosure Act—approximately $5 million. Sorry, how many centimeters of the Keystone pipeline have been built? Can you help me on that one?[6]

The "revolution" Levick was referring to was the upheaval in the nature of information, and especially the power it gave ordinary people to participate in public affairs. Industrial actors are wedded to a "pre-revolutionary" style of communication, he lamented; an old-school, republican form of communication—"small 'R,' as in a republic"—in which the speakers maintain control over the narrative. The interactive, dialogic, authoritative nature of information in the contemporary media ecosystem gave power to those who didn't have the financial resources or the "facts," as Levick put it, but could capture the symbolic resources needed to push back against everything the industry stood for. Despite its long legacy of economic and political advantage, Levick insisted, the industry had lost control over the ability to define the problems facing the public.

"We are now talking about a democracy!" he cried, as the attendees shifted in their seats. "Where are we in that messaging? How are we communicating?" The task at hand, Levick insisted, was to get out in front of a problem before it became a problem. To see an issue simmering, and get it off the heat before it boiled over:

> How do ideas become movements? First it's talked about in Ridgewood [a small American town]. Very few people are conversing about it. Then it becomes slowly more and more popular. *That* is the moment when we begin to influence, before minds are made up.

"Truth is not about the facts. Truth is what we know first!" Levick added, to sighs of understanding from the room. The professional communicator's job is to create this truth: to define the problem in one's own terms, to engage with the publics that matter, to situate your message in their context of understanding and provide reasons and arguments to persuade them your view is valid. *This* is the role of communication in a democracy, Levick claimed. Publicity is the tool by which democratic publics and their problems are given shape and made meaningful, or contained and dispersed.

Throughout the two-day conference, during the panels, lunches, receptions, and working group sessions, the imperative for concerned publics to be engaged and to engage others in the resolution of public problems was a central theme. We heard from representatives of the House and Senate, regional trade associations, energy and environmental policy advisers, the US Chamber of Commerce, and public relations and public affairs directors in multiple sectors. David Holt, president of the Consumer Energy Alliance, a trade and lobby group with around 280 corporate members and thousands of

individual supporters, emphasized the need for attendees to develop "aspira-
tional" communications for their publics:

> Not only do we need to motivate our base and make them turn out and
> make them aware that jobs and the future of the nation are at stake, but we
> need to try to find ways to inspire and elevate that conversation. Then it
> makes it much easier to motivate.[7]

To inspire and elevate, to bring people together around the issues at stake,
conference presenters described the repertoire of skills and techniques
drawn on by the professional communications strategist. These skills and
techniques will be familiar to the reader: they appear in these pages as the
product of the last 100 years of PR's progress. Creating coalitions of support
across state- and local-level organizations; mobilizing third-party "grassroots" advocates, such as employees in your organization, to speak on your
behalf; crafting data points and statistics to factualize persuasive narratives;
extensive media monitoring and tracking of opposing groups' public presence; pro-energy and economic growth (and anti-regulatory) information
and influence campaigns; scenario planning to anticipate problems before
they start; public events designed for promotional purposes.[8]

The need for this strategic nature, the conference speakers insisted, was to
bring us together around what it is we care about as citizens in a democratic
society. Just as Nader had done forty years ago, industrial actors could gain
adherents to their cause by finding common ground, stabilizing the territory
on which public purpose could be found. Beyond the rational arguments,
the finding of facts, and the critical debates stood a moral obligation to co-
here around what matters, to find the conditions of compromise by which
everyone could agree. For Levick, Holt, and the others, this terrain was a collective commitment to the health and safety of the environment. This was
the commitment to which we could all aspire, overcoming the "us versus
them" framing of environmentalism against energy production. "Every
single person in this room is an environmentalist," Holt concluded. "We are
all environmentalists."[9]

Before we dismiss this claim outright as the cynical maneuvering of a
corporate shill or uphold it as a pragmatic position in an industrial energy-
dependent society, it is worth considering how this rhetoric operates to
constitute a notion of publics and their problems. At base, the idea that
"everyone" is concerned about the same issue, that "we all" care about the

environment, and that popular decision is required to establish the environment as a matter of public concern—these are in themselves laudable goals within a participatory democracy. We need not observe the machinations of the oil and gas industry for long, however, before recognizing that the "we" is not meant to include all citizens, let alone all those affected by the industry's degradation of environmental health; and that environmentalism is not a stable concept but a compromise object in and of itself, used to make things "sound right" to all concerned while legitimating practices that do nothing to preserve or protect the global ecosystem or to mitigate the climate crisis we currently face.

If we expand the aperture of the lens, moving out from a focus on the politics to encompass the publics who are meant to "engage" with political problems, we can better analyze the central role of public relations in American life. The true struggle seems to be lodged in the ability to define what "the public" means. In the articulation of common ground we are thinking of who may stand on this ground; and the continued emphasis on communication that encourages participation implies that the more people who come out and participate, the more robust and powerful the public. This is clearly the logic animating the practice of public relations. In its campaigns to inform and influence, its coalitions and networks, its plans for representation, its lobbies and allies, PR seeks to create majority publics who will provide consent for the project at hand.

The purpose of strategic communication is therefore to devise the rationales and incentives to persuade members of the public to engage. This is not merely about presenting the "facts" of the matter; it is about gathering up people's concerns, of connecting to the things they care about. Most of us are familiar with the contours of this approach: persuasive appeals must capture not only the minds but also the hearts of the audience to be effective.

Public relations is nothing less than the professionalization of public-making. And in this sense, we might think its task is to produce discourses that address members of the public as a concerned public, presenting the social and political concerns of the day in a manner that engages them to take action, to debate the various sides of the affair, to come to reasonable compromises or consensus, to stand on common ground.

But this is not what takes place. The true measure of a successful public relations campaign is the extent to which it has ensured that publics do *not* form, do *not* constitute a body of concern, and do *not* raise problems as public problems. Whether in its short-term mode (e.g., crisis communications) or its long-term strategies (e.g., issue management), public relations exists to *control* the way citizens come together to see themselves as members of a legitimate public and to recognize an issue as a legitimate problem. This is the essence of PR as a technology of legitimacy: to mediate publics and problems so that they can appear or disappear in political contexts of importance.

Public relations creates, shapes, and promotes a politics that is embedded in our major institutions, our common practices of mediated debate, and the way we collectively think about what "the public" is and what it ought to do.[10] This conception of democratic politics is so deeply embedded in our habits of action that even when we fight for better representation of those voices that are continually left unheard or denied participation or the right to engage, we retain its premises rather than attempting to challenge it at its base. We turn to publicity to inform, engage, and mobilize. We seek out like-minded supporters, reduce issues to their essence, and create antagonists to shore up our own boundaries of who is inside or outside of "our" concerns and values.

In many instances, including those described in this book, public participation ("engagement") and increased opportunities for deliberation have failed to amount to democratization, reinforcing rather than overcoming historic inequalities and maintaining the legitimacy of existing structures of authority.[11] This has historically been patterned by corporate and state interests, but it is not limited to these. Professionals in nonprofit and nongovernmental sectors are also invested in maintaining the political armatures of participation, deliberation, and compromise—if not for ideological reasons, then for practical ones.

The charismatic politics of the PR figures we have encountered in these pages (as well as the charisma of their data and infrastructural mediations)[12] is predicated on appeals to self-interest, immediate situations, and directly implicated concern. This is the "stakeholder" model of public formation: one that relies on participation and engagement of directly affected parties rather than the established "truths" of inconvenient facts in the realm of science or politics. A stakeholder model of publics is built around the notion of risk. Risks, especially the risks of publicity, can appear in a variety of forms: the risks of lack of public trust or confidence in state or corporate representatives; the risks of public calls for transparency in major social and political institutions; the risks that information technologies allow its users to circulate multiple perspectives. Appealing to stakeholders distributes risk among a range of "decision-makers" whose participation stabilizes and renders

more incontrovertible the outcome of debate. It establishes a ground of consensus and compromise that operates beyond scientific or economic data. It
appears more legitimate and representative of social values than do the insistent claims of scientists. It is oriented to process rather than to the quality
of scientific analysis.[13]

Most important, the stakeholder model of decision-making allows the facilitator of this model to determine what problems go before the public and
how they are framed as problems. When Levick insists his clients "engage"
with their audiences, when Holt claims that we are all environmentalists,
these are expressions of how clients can take control; how they can "manage"
their publics to prevent the formulation of problems that will act back on their
instigators. This is the "relations" part of public relations: posing problems
as contests of legitimacy among competing stakeholders whose shared communication will frame and resolve their problematic nature, protecting the
true owner of the problem from full accountability.

The scholar Chris Russill has observed that the emphasis on "'communicative' conceptions of democracy" rooted in deliberative, participative, and
conversational models of public life has left behind earlier understandings
of "the authority of scientific models of inquiry in the fields of culture and
politics."[14] He proposes that we return to a spirit of inquiry—the recognition
of a problem as a problem—that lies at the heart of John Dewey's and Walter
Lippmann's early twentieth-century theories of the formation of publics.

Inquiry is an idea that takes the interdependency of self and other as a precondition for the formulation of problems and the organization of publics
around them. It is not about establishing direct lines of self-interest into
an immediate object of concern but about a holistic conception of distant
troubles and felt anxieties as part of close-up consequences. These troubles
and concerns disrupt our established patterns of conduct, change our expectations of how the world works, and demand attention as social problems: "It
is inquiry—the active shaping of difficulties and felt concerns as problematic
situations—that brings publics into existence."[15]

This "problem-responsive" account of inquiry seems especially germane
to our relationship to the environment. The environment is not a problem
of politics; nor is it a problem of publicity. But in making it appear that way
over the last hundred years, we have turned it into something that seems to
require political solutions, wielding the techniques of democracy we have
at our disposal. These techniques, however, offer a model of the citizen that
allows the pretense of communication, participation, and "engagement" to

substitute for the deep awareness of the environment as an interdependent system in which our actions affect the actions of others. When it comes to the environment, relying on a model of problem formulation built on controlled participation, short-term fixes, and resolvable issues will always leave intact the true nature of the problem: to create a collective sense of concern, to come to terms with the obstacles to our continued existence on the planet.

The "information and influence campaign" needed now is not one that allows everyone "to get what they want" in public affairs but to formulate the problem as one that truly affects everyone, no matter how distant or unseen. This is not a matter of crafting a more persuasive narrative, engineering a more informed debate, or developing better data. It is a problem of rethinking the relationships among our cherished concepts and their opponents, of breaking down barriers between "us" and "them," expert and citizen, society and nature, past and future, facts and felt truths, in the articulation of what matters. That is the strategic nature we need.

What would it take to work toward this kind of campaign? In tracing the twin evolution of public relations and environmentalism in the last 100-plus years, one sees the many relations created among environment, information, and publicity; but this environmentalism is also full of cracks and empty spaces; of places, people, and problems left behind; of decisions designed to exclude; of expert knowledge crafted at the expense of existential concerns; and of extractive techniques—both material and symbolic—made legitimate through neglect of human health and nature's balance. It is in bringing these absences to light that we can begin to reconstruct the environment as a matter of concern by which we are all truly affected and so can privilege as the ultimate public problem.

Interviews and Observation Sites

Interviews

350.org
B-Team
Brigham Young University
Burson-Marsteller
Caplan Communications
Center for International Earth Science Information Network
Centre for Internet and Society
CEPEI Colombia
Climate Nexus
Cornell Institute for Climate Change and Agriculture
Crimson Hexagon
Earth Networks
Environmental Defense Fund
Dalberg Data Insights
DG + CO
Dow Chemical Company
DTN
Edelman
FKHealth
FSG
FTI Consulting
Global Call for Climate Action
Global Strategic Communications Council
Hoggan & Associates
Humanitarian OpenStreetMap
Kekst CNC
Interel Belgium
LIRNEasia
M + R
National Audubon Society
O'Dwyer's
Ogilvy PR
Penn State University

Procter & Gamble
Public Relations Society of America
Pulse Lab Jakarta
Qorvis Communications
Rockefeller Foundation
Sanchis & Associates, Spain
Skoll Global Threats Fund
SMK Netherlands
Spector Associates
Sustainable Energy for All
TE Connectivity
TÜV SÜD
United Nations Global Pulse
United Nations Population Fund
World Environment Center
World Economic Forum
Yale Program on Climate Change Communication
Zero to Sixty Communications
+ independent consultants

Observation Sites

2017   American Climate Leadership Summit, Washington, DC
2017   Public Relations Society of America Annual Meeting, Boston, MA
2018   The Communications Network (ComNet) Annual Conference, San Francisco, CA
2018   Data for Good Exchange (D4GX) Bloomberg, New York, NY
2018   Epidemic Readiness and Trustworthy Data Workshop (webinar), World Economic Forum
2019   Analytics Day: Data for Good, Canadian Imperial Bank of Commerce, Toronto, Canada

E. Bruce Harrison Company, List of Clients, 1973–1997

This list, a compilation of four company client rosters, client case histories prepared by the E. Bruce Harrison Company, and a trade publication article, needs to be read against the grain.* Some of these clients are associations or coalitions formed by Harrison himself (e.g., the National Environmental Development Association [see chapter 3]); some are companies in which he used to hold positions (e.g., he was vice-president at Freeport Minerals [now Freeport-McMoRan] in the late 1960s before starting his own firm, which then represented Freeport); and some are companies with which he formed alliances to serve different industries (e.g., in 1982 Harrison was on retainer to Glick & Lorwin, Inc., a PR firm in New York City, as its "Washington presence"). Coalitions listed in the second column were also clients of Harrison. In other words, Harrison represented and provided services for the coalition as a unit in addition to performing work for individual company clients who may have been members of those coalitions.

It is also not always clear what kinds of work Harrison performed for these clients. In some cases, there are extensive and long-standing connections and multiple efforts to sidestep federal environmental regulation for clients or offset negative media coverage about them. This can be discerned by the ongoing participation by companies in different coalitions Harrison organized (among other kinds of participation, as documented in this book). In other cases, one-time services were provided. The notes in the right-hand column are taken directly from listed sources and therefore do not reflect the scale or scope of the services. Still, the list gives a general sense of the breadth of Harrison's influence across American industries over a determinate time period (1973–1997) and of the sweeping range of companies, organizations, and sectors seeking publicity through specialized "green" public relations during this period.

Client (Source in brackets)	Type of Service or Coalition Membership
Adolph Coors Company (2) (3) (4) (6)	Employee relations; grassroots
A. E. Staley, Inc. (1)	Electricity Consumers Resource Council
Airco Educational Services (1)	Electricity Consumers Resource Council
Airco Industrial Gases (1)	Company client
Air Conditioning Contractors of North America (1)	Company client
Air-Conditioning and Refrigeration Institute (1) (5)	Company client
Air Products and Chemicals, Inc. (1) (2) (6)	Electricity Consumers Resource Council; environmental, health & safety communication; risk communication; media relations
Allied Corporation (1) [Allied-Signal (6)]	NEDA-Groundwater; NEDA-CWP
Alpha Twenty-One Corporation (1)	NEDA
Alternative Materials Institute (4)	Company client (coalition organized by Harrison); organizational management
Aluminum Company of America [ALCOA] (1)	Process Gas Consumers Group
American Association of State Highways & Transportation Officials (1)	NEDA
American Automobile Association (3)	Company client
American Automobile Manufacturers Association (3)	Company client
American Can Company (1)	Process Gas Consumers Group
American Ceramic Society (3) (6)	Company client; legislative monitoring; marketing
American Express (2) (6)	Community relations
American Meat Institute (1)	Coalition for Food Irradiation
American Medical Association (3) (6)	Company client; animal rights communications program
American Medical Laboratories, Inc. (6)	Environmental communications; media relations
American Petroleum Institute (3)	Company client
American Sugar Alliance (4)	Public affairs/grassroots communication program

Client (Source in brackets)	Type of Service or Coalition Membership
American Textile Manufacturers (2)	Environmental policy
AMFAC (3) (6)	Company client
AMREP Corporation (1)	NEDA
Anheuser-Busch Companies, Inc. (1) (2) (3) (6)	Electricity Consumers Resource Council; NEDA-CWP; environmental, health & safety communication; legislative monitoring
Annapolis Center for Environmental Quality (3)	Company client
Aristech Chemical Corporation (USX) (2) (3) (4) (6)	Environmental policy; community relations; crisis management; risk communication
Armco Inc. (1)	Electricity Consumers Resource Council; Process Gas Consumers Group
ASB Capital Management (6)	Company client; media relations
Asea Brown Boveri (6)	Marketing
Aseptic Packaging Council (3)	Company client
Ashland Oil, Inc. (1) (6)	NEDA; NEDA-CAAP
AT&T (2) (3) (6)	NEDA; environmental, health & safety communication; environmental policy; employee communication; marketing
Autochoice (later renamed Coalition for Vehicle Choice) (6)	Coalition; grassroots; media relations
Atlantic Richfield Company (ARCO) (1) (3)	NEDA; NEDA-CAAP; NEDA-Groundwater; NEDA-CWP; Process Gas Consumers Group
AZS Corporation (6)	Crisis management
BASF (2)	Environmental, health & safety communication; crisis management; risk communication
Bethlehem Steel Corporation (1)	Electricity Consumers Resource Council; Process Gas Consumers Group
Big B Ranch (1)	NEDA
Billy Rogers Farm (1)	NEDA
Blockbuster Entertainment (3) (6)	Company client
Blue Cross/Blue Shield of Virginia (3) (6)	Company client
Bombardier, Inc (4)	Environmental awareness campaign
Booz-Allen & Hamilton (1)	Company client
Borg-Warner Corporation (1)	Process Gas Consumers Group

Client (Source in brackets)	Type of Service or Coalition Membership
BP America (2) (3) (4)	Environmental, health & safety communication; environmental policy; benchmark studies
Braken, E. O. (1)	NEDA
Bryan Landfill (3) (6)	Company client
Building and Construction Trades Department, AFL CIO (1)	NEDA; NEDA-CAAP; NEDA-CWP
Burlington Industries, Inc. (1)	Process Gas Consumers Group
Business Roundtable (1) (4)	Company client; public affairs; media relations
CAE (3)	Company client
Campbell Soup Company (1)	Coalition for Food Irradiation; NEDA; NEDA-Groundwater; NEDA-CWP
Canal Barge Company, Inc. (1)	NEDA
Capital Yacht Club (1)	Company client
ChemGen (3) (6)	Company client
Chemical Manufacturers Association (2)	Environmental, health & safety communication
Chevron U.S.A. (1)	NEDA-CAAP; NEDA-Groundwater; NEDA-CWP
Chrysler Corporation (1) (3) (6)	Process Gas Consumers Group; grassroots; media relations
Cincinnati Gas & Electric Company (1)	NEDA
Citibank (2)	Marketing communication
Clairol, Inc. (1) (2) (6)	Company client; community relations; media relations
Clean Air Act Project (1)	Coalition; public awareness campaign
Clean Water Project (1)	Coalition
Clorox Company (2) (3) (4) [Javex: 6]	Environmental, health & safety communication; marketing communication; crisis management; recycling media program; international communication
Coalition for Vehicle Choice (3) (4) (6)	Coalition; public affairs/grassroots; environmental communications
Coca-Cola USA (2)	Environmental, health & safety communication; marketing communication; recycling media program
Colgate-Palmolive (2) (4) (6)	Environmental, health & safety communication; employee communication; corporate environmental policy; global marketing plan; international monitoring/analysis

Client (Source in brackets)	Type of Service or Coalition Membership
Computer Technologies Corporation (1)	Process Gas Consumers Group
Cone Mills Corporation (1)	Electricity Consumers Resource Council
Consolidation Coal Company (1)	Company client; NEDA-CAAP
Corning Glass Works (1)	Electricity Consumers Resource Council; Process Gas Consumers Group
Cosmair (2)	Environmental, health & safety communication
Cosmetics, Toiletry & Fragrance Association (2) (3) (4) (6)	Environmental, health & safety communication; animal rights; coalition formation; grassroots
Council of Former Governors (1)	Company client
CP Chemicals (1)	Company client
CSC Logic ,Inc. (6)	Marketing
Dallas Housing Authority (3)	Company client
Dallas International Sports Commission (6)	Community relations; media relations
Del Monte Corporation (1)	Coalition for Food Irradiation
Destec Energy (Dow Chemical Company subsidiary) (3) (4) (6)	Company client; survey, media support; marketing
Diamond Shamrock Chemical Company (1)	Electricity Consumers Resource Council
Dow Chemical Company (1)	Electricity Consumers Resource Council; NEDA-CAAP
Du Pont (2)	Community relations
Eaton Corporation (1)	Process Gas Consumers Group
Edison Electric Institute (2)	Environmental, health & safety communication
E. I. du Pont de Nemours, Inc. (1) (4) (6)	Company client; community relations; Coalition for Food Irradiation; Electricity Consumers Resource Council
Electric Power Research Institute (3) (6)	Company client
Electric Vehicle Council (1)	Company client
Electricity Consumers Resource Council (1) (5) (6)	Company client; legislative monitoring; media relations
Englehard Corporation (1) (3) (6)	Sorptive Minerals Institute
Excel-Minerals Company (1)	Sorptive Minerals Institute
Exxon Company, USA (1)	NEDA; NEDA-Groundwater
Federal Maritime Commission (1)	Company client

Client (Source in brackets)	Type of Service or Coalition Membership
Figgie International (6)	grassroots
Florida Fruit & Vegetable Association (1)	NEDA; NEDA-CAAP; NEDA-CWP
Florida Power and Light Company (1)	NEDA; crisis management
Florida Sugar Cane League (1)	NEDA
Florida Water Users Association (1)	NEDA
Floridin Company (1)	Sorptive Minerals Institute
Fluor Corporation (1)	NEDA-CAAP
FMC Corporation (1)	Company client; NEDA
Ford Motor Company (1) (3) (6)	Process Gas Consumers Group; grassroots; media relations
Fred Harvey Company (4)	Corporate environmental report preparation and promotion
Freeport Minerals Company (1)	Company client
Frito-Lay Foods (2) (6)	Community relations
Garrison Diversion Conservancy District (1)	NEDA
Gates Energy Products (6)	Marketing; media relations
Gerber Foods, Inc. (1)	Coalition for Food Irradiation
General Dynamics (2) (4)	Environmental policy; community relations
General Electric Company (1)	NEDA-Groundwater
General Foods Inc. (1)	Coalition for Food Irradiation
General Mills, Inc. (1)	NEDA-CWP
General Motors Corporation (1) (3) (6)	Company client; Electricity Consumers Resource Council; NEDA; NEDA-CAAP; NEDA-CWP; Process Gas Consumers Group; grassroots
General Signal Company (2) (3) (4) (6)	Environmental, health & safety communication; strategy development
Georgia Tennessee Mining and Chemical Company (1)	Sorptive Minerals Institute
Glick & Lorwin, Inc. (1)	Company client
Global Climate Coalition (3) (4) (6)	Company client; media relations campaign
Good Water, America (1)	Company client
Gulf Coast Waste Disposal Authority (1)	NEDA

Client (Source in brackets)	Type of Service or Coalition Membership
Gulf Oil Corporation (1)	NEDA
Halogenated Solvents Industry Alliance (1) (4)	Company client; grassroots communication program
Hartz Mountain Corporation (1)	Sorptive Minerals Institute
Hawkins Ranch (1)	NEDA
Hechinger Stores (2)	Employee communication; marketing communication
Hercules Incorporated (1)	Electricity Consumers Resource Council
Hershey Foods Corporation (1)	Electricity Consumers Resource Council ; Process Gas Consumers Group
Highway Users Federation (4)	Public affairs; alliance formation; media relations
Hoechst Celanese (2) (3) (6)	Environmental, health & safety communication; employee relations; community relations; crisis management; risk communication; public-interest counsel; media relations
Hoffman-LaRoche (3) (6)	Company client
Honeywell (1)	NEDA
Honor Guard Security Services (1)	Company client
Houston Natural Gas Corporation (1)	NEDA
IBM Corporation (1)	NEDA-CAAP; NEDA-Groundwater
ICMA Retirement Corporation (6)	Marketing
Industry Coalition for Fire Safety (1)	Company client
Industry Cooperative for Ozone Layer Protection (ICOLP) (4) (6)	Strategic direction, media relations, organizational management, daily operations; international monitoring/analysis
Institute of [for] Resource Recovery (4)	Grassroots network organizing and activation
International Association of Bridge, Structural and Ornamental Iron Workers (1)	NEDA
International Bottled Water Association (6)	Marketing
International Brotherhood of Electrical Workers (1)	NEDA; NEDA-CWP
International Hardwood Products Association (6)	Media relations

Client (Source in brackets)	Type of Service or Coalition Membership
International Union of Operating Engineers (1)	NEDA; NEDA-CWP
International Year of Disabled Persons (1)	Company client
I. V. Duncan Ranch (1)	NEDA
Join Hands (3)	Company client
Kaiser Aluminum & Chemicals Corporation (1)	Electricity Consumers Resource Council; NEDA; NEDA-CAAP; NEDA-CWP
Keystone Consolidated Industries, Inc. (3) (6)	Company client
King Ranch (1)	NEDA
Koppers Industries (2) (3) (4) (6)	Environmental, health & safety communication; crisis management; media relations
Kraft, Inc. (1)	Coalition for Food Irradiation
Laborers' International Union of North America (1)	NEDA; NEDA-CWP
Laidlaw Waste Systems (3) (6)	Company client; community relations; environmental communications; media relations
Lake County Forest Preserve District (4)	Animal rights; communication program
Las Colinas Landscape Services (6)	Employee relations
Lowe's, Inc. (1)	Sorptive Minerals Institute
LTV Steel Company (1)	Process Gas Consumers Group
Maxus Energy (3) (6)	Company client
McCormick and Company, Inc. (1)	Coalition for Food Irradiation
McDonald's Corporation (4)	Solid waste management program
McKenna & Cuneo (3) (6)	Company client
Merco Joint Venture (3) (6)	Company client; crisis management; community relations; environmental communication; grassroots; international monitoring/analysis; legislative monitoring; marketing; media relations
Metropolitan Police District of Columbia Vest Fund (1)	Company client
Metro Washington Home Improvement Council (1)	Company client
MITRE Corporation (1)	Company client
Mitsubishi (2)	Environmental, health & safety communication; community relations; crisis management

Client (Source in brackets)	Type of Service or Coalition Membership
Mobil Oil Corporation (1) (3) (4)	NEDA; NEDA-CAAP; NEDA-Groundwater; benchmark study
Monsanto/Vista Chemical (2) (3) (4) (6)	Environmental, health & safety communication; research, worldwide monitoring, communication support activities; public affairs; legislative support; grassroots; media relations
National Agricultural Chemicals Association (4) (6)	Public affairs; media relations; environmental communications
National Association of Manufacturers (1) (5)	Company client
National Association of Private Psychiatric Hospitals (1)	Company client
National Cattlemen's Association (1)	NEDA
National Council of Agricultural Employers (1)	Company client
National Environmental Development Association (1) (3) (5)	Coalition
National Food Processors Association (1)	Company client; Coalition for Food Irradiation
Natural Gas Consumers Information Center (1)	Company client
National Home Improvement Council (1)	Company client
National Marine Services, Inc. (1)	NEDA
National Medical Enterprises (3) (6)	Company client
National Pork Producers Council (1)	Coalition for Food Irradiation
National Realty Committee (1)	Company client
National Science Foundation (5)	Company client
National Solid Wastes Management Association (4)	Media training
National Waterways Conference (1) (5)	Company client
National Women's Economic Alliance* (1) (3) (6)	Company client
New Orleans Public Service, Inc. (1)	NEDA

Client (Source in brackets)	Type of Service or Coalition Membership
North Texas Cement Company (3) (6)	Company client
North Texas Commission (4)	Coordinating proposal for Superconducting Super Collider; media relations
Norton Company (6)	Media relations
NovaCare, Inc. (3) (6)	Company client
Occidental Petroleum Corporation (1)	NEDA-CAAP; NEDA-Groundwater; NEDA-CWP
Oil-Dri Corporation of America (1)	Sorptive Minerals Institute
Olin Corporation (1)	Electricity Consumers Resource Council
Operating Industries, Inc. Landfill (4)	Community, government and media relations campaign
Owens-Corning Fiberglas Corporation (1)	Process Gas Consumers Group
Owens-Illinois, Inc. (1)	Process Gas Consumers Group
Pennzoil Company (1)	Company client; NEDA; NEDA-CWP
Pfizer Pharmaceuticals (2) (6)	Environmental, health & safety communication; crisis management; grassroots
Phelps Dodge Corporation (3)	Company client
Philip Morris (2) (3)	Environmental policy
Phillips Petroleum Company (1) (3) (6)	Company client; NEDA; NEDA-CAAP; NEDA-Groundwater; NEDA-CWP; international monitoring/analysis
Port of Port Angeles (1)	NEDA
PPG Industries [Pittsburgh Plate Glass Company] (1) (6)	NEDA; NEDA-CAAP; NEDA-CWP; international monitoring/analysis
Process Gas Consumers Group (1) (3) (6)	Company client; media relations
Procter & Gamble Company (1)	NEDA-CAAP
Pro-Trade Group (4)	Media support
Public Environmental Reporting Initiative (3) (6)	Company client
Public Service Company of Indiana (1)	NEDA
Public Service Company of New Mexico (1)	NEDA
Ralston Purina Company (1)	Coalition for Food Irradiation

Client (Source in brackets)	Type of Service or Coalition Membership
Rhône-Poulenc (2) (3) (6)	Company client; environmental, health & safety communication; community and employee relations; risk communication; media relations
R. J. Reynolds/Nabisco** (3) (6)	Company client
Rochester-Pittsburgh Coal Company (1)	NEDA
Salomon Inc. (3)	Company client
Salt River Project (1)	NEDA
Sandoz (2)	Environmental, health & safety communication
Santa Clara Landfill Coalition (3) (6)	Company client
SEED (1)	NEDA
Seifman, Semo & Slevin (1)	Company client
Sherman Wire (3) (6)	Company client
Smokeless Tobacco Council (6)	Media relations
Society of National Association Publications (1)	Company client
Society of the Plastics Industry (1)	Company client
Sonat Marine, Inc. (1)	NEDA
Sorptive Minerals Institute (1) (3) (4) (5)	Company client; federal affairs; organizational management
Southern (Power) Company (2)	Environmental policy
Standard Oil Company (Indiana) (1)	NEDA-CAAP
Standard Oil Company (Ohio) (1)	NEDA-CAAP; NEDA-Groundwater; NEDA-CWP
Stauffer Chemical Company (1)	Electricity Consumers Resource Council
Sterling Winthrop (3) (6)	Company client
Sumitomo Bank, Ltd. (6)	Legislative monitoring
Sugar Cane Growers Cooperative of Florida (1)	NEDA
Sun Company, Inc. (1)	NEDA; NEDA-CAAP; NEDA-Groundwater; NEDA-CWP
Tenneco, Inc. (1) (2)	NEDA; NEDA-CAAP; NEDA-CWP; environmental, health & safety information; environmental policy
Texaco, Inc. (1)	NEDA; NEDA-CAAP

Client (Source in brackets)	Type of Service or Coalition Membership
Texas-New Mexico Power Company (3) (6)	Company client; community and employee relations; environmental communications; legislative monitoring; marketing; media relations
Total Indoor Environmental Quality Coalition (TIEQ) (3) (4) (6)	Company client; coalition formed by Harrison
Trane Company (3) (6)	Company client
Tri-City Health Center (3) (6)	Company client
Trinity River Authority (1)	NEDA
Union Carbide Corporation (1) (3)	Company client; Electricity Consumers Resource Council
Union Oil Company of California (1)	NEDA
Uniroyal Chemical (2) (3) (4) (6)	Environmental, health & safety communication; environmental policy; government, community and employee relations; crisis management
United Association of Journeymen & Apprentices of the Plumbing and Pipe Fitting Industry (1)	NEDA; NEDA-CWP
United Brotherhood of Carpenters and Joiners (1)	NEDA; NEDA-CWP
United Fresh Fruit and Vegetable Association (1)	Company client
United Union of Roofers, Waterproofers & Allied Workers (1)	NEDA
UNOCAL (3) (6)	Company client
US Advanced Ceramics Association (4)	Trade association incorporated and developed by Harrison; organizational management
US Agency for International Development (3) (6)	Company client
US Army (2)	Risk communication
US Department of Commerce (1)	Company client
US Department of Energy (1) (5)	Company client
US Environmental Protection Agency (3) (6)	Company client; environmental communications
US Ecology (3) (6)	Company client
US Steel Corporation (1)	Electricity Consumers Resource Council
US Sugar Corporation (1)	NEDA

Client (Source in brackets)	Type of Service or Coalition Membership
Velcon Filters, Inc. (1)	NEDA
Velspar Paints (3) (6)	Company client
Waste Management [of North America], Inc. (2) (6)	Environmental, health & safety communication; community relations; grassroots; legislative monitoring; media relations
Waverly Mineral Products Company (1)	Sorptive Minerals Institute
Welder, Leo (1)	NEDA
Westcott Communications (6)	Marketing; media relations
Western Union (3) (6)	Company client
Westvaco Corporation (1)	NEDA; NEDA-CAAP
Weyerhaeuser Company (1)	NEDA-CAAP
Whitman and Ransom (6)	Crisis management
Wittenburg [sic; possible Whittenburg] J. A. III (1)	NEDA
Wood, R. L. (1)	NEDA
Wooten, Frank, Jr. (1)	NEDA
Zexel Corporation (3) (6)	Company client
Zoecon Corporation (2) (3) (6)	Community and employee relations; partnerships with local public interests; crisis management; marketing; media relations

*1. E. Bruce Harrison Company: Company & Coalition Clients (n.d.)
 2. E. Bruce Harrison, Summary of Client Engagements: 1987–1997
 3. E. Bruce Harrison Company, The Sustainable Communication Company: Harrison Clients (n.d.)
 4. E. Bruce Harrison Company, Case History Index and Case Histories (n.d.)
 5. "D. C. Agency Created First Client," *Publicist*, March/April 1982.
 6. E. Bruce Harrison Company, Client Services; Coalition and Association Clients (n.d.)
** The NWEA was created by Patricia de Stacy Harrison in 1983, ten years after she and her husband, E. Bruce Harrison, had co-founded the Harrison Associates public relations firm (Harrison & Associates would be renamed the E. Bruce Harrison Company in 1978). Source (1) lists corporate sponsors of the National Women's Economic Alliance (NWEA) Foundation as Harrison clients. However, we have no evidence that these companies sought public relations representation by EBH. While NWEA is included as a client of Harrison's, therefore, the corporate sponsors of that association are not included in this master list.
*** RJR Nabisco was formed in 1985 through the merger of the R. J. Reynolds Tobacco Company and Nabisco Brands food products. In 1999, in the wake of a major class action lawsuit against Big Tobacco, the R. J. Reynolds tobacco business was spun off again into a separate company.

Notes

Introduction

1. Hertsgaard and Pope, "Fixing the Media's Climate Failure," 12.
2. Bill McKibben, "Covering Climate Change," Columbia Journalism Review (public event), 30 April 2019. https://www.youtube.com/watch?v=FO9DKk07SCY&t=1498s.
3. See Oreskes and Conway, *Merchants of Doubt*; Miller and Dinan, *A Century of Spin*; Beder, *Global Spin*; Coll, *Private Empire*; Hoggan, *Climate Cover-Up*; Mayer, *Dark Money*; Stauber & Rampton, *Toxic Sludge Is Good for You!*; Pooley, *Climate War*; Silverstein, *The Secret World of Oil*; Gelbspan, *The Heat Is On* and *Boiling Point*.
4. Hertsgaard and Pope, "Fixing the Media's Climate Failure," 14.
5. Dewey, *The Public and Its Problems*, chapter 5, "Search for the Great Community," 143–184. Dewey also worried about conditions that favor the overspecialization of knowledge, the static hold on outdated traditions, and the sowing of partisan beliefs.
6. Lippmann, *Public Opinion*, 367. Stuart Ewen argues that this view of informed public individuals "was eloquently expressed by Thomas Jefferson in his second inaugural address, when he declared that the 'diffusion of information and the arraignment of all abuses at the bar of public reason, should be the creed of our political faith—the text of civil instruction.'" For Ewen, "intrinsic, here, was the assumption that democracy depended on the existence of a literate middle-class public, apprised of current events, continually engaged in discussion." Ewen, *PR! A Social History of Spin*, 50.
7. The ideas of John Dewey and Walter Lippmann have structured American theories and research on mass communication. John Durham Peters has argued that Dewey and Lippmann (along with sociologist Paul Lazarsfeld and political scientist and public policy scholar Harold Lasswell) treated "the public" as a key problem for social science. Mass communication, Peters points out, has always been about the possibilities and limitations of democracy. "A political concern for democracy is thus not only a *topic* of discourse in American mass communication theory; it is part of the *structure* of that discourse." Peters, "Democracy and American Mass Communication Theory: Dewey, Lippmann, Lazarsfeld," 200.
8. Craig Calhoun summarizes Habermas's criticism of public opinion management as the "staged display" of publicity rather than an organized process of consensus formation: "Public-opinion research is more akin to the simultaneously developed field of group psychology than to democratic practice; it is an auxiliary science to public administration rather than a basis or substitute for true public discourse." Calhoun, ed., *Habermas and the Public Sphere*, 29. See also Bourdieu, "Opinion Polls: A Science Without a Scientist." At the Canadian Energy Summit in Toronto in 2015, a well-known pollster called public opinion research the "gamification of opinion," referring

to the idea that opinion-making is a strategic battleground, won or lost not by what anyone actually thinks or does, but by the protagonist's ability to generate the appearance of a popular viewpoint.

9. See Sheingate, *Building a Business of Politics*, 50–65, for a longer discussion of Bernays's claims to expertise through his associations with social scientists. See also Ewen, *PR!* (chapter 8) on Bernays's transformation of Lippmann's views to promote public relations practitioners as public information specialists.

10. Bernays, "Manipulating Public Opinion: The Why and the How," 961.

11. Jansen, "Semantic Tyranny." See also Schudson, "The 'Lippmann-Dewey Debate' and the Invention of Walter Lippmann as an Anti-Democrat, 1986–1996"; and Schudson, "Walter Lippmann's Ghost: An Interview," 31–40.

12. Rabin-Havt, *Lies, Incorporated*.

13. Arendt, "Truth and Politics," 54.

14. Suchman, "Managing Legitimacy." See also Davidson and Gismondi, *Challenging Legitimacy at the Precipice of Energy Calamity*.

15. Lee Edwards argues that organizational PR appears to create "pathologies of deliberation" because it contravenes the conditions of deliberative democracy: it is driven primarily by self-interest, targets specific and not broad audiences, and uses persuasive instead of rational communication to gain power and influence. She proposes a broader interpretation of public relations as maintaining a deliberative capacity that can be deployed for democratic purposes "depending on who is using it and what they are using it for." Edwards, "The Role of Public Relations in Deliberative Systems," 74. We take this viewpoint into a slightly different direction: the extensive organizational and institutional power over systems of communication and information means that the democracy we have is inbuilt with these so-called pathologies. Following Elisabeth Clemens, our approach "does not deny that politics may be driven by self-interest but asks how 'self-interest' is constructed and under what conditions it becomes the dominant script guiding political action." Clemens, *The People's Lobby*, 9. Evaluating PR as practice is helpful to avoid stale reification or critique of solely top-down initiatives; but it leaves intact some of its most problematic aspects, such as what information is made valuable and how; justification of means to a desired end; and severe resource differentials among different organizational actors.

16. Edwards, *Understanding Public Relations*, chapter 2.

17. Edwards, *Understanding Public Relations*, 5. See also Ewen, *PR!*, 33.

18. Haas, *Epistemic Communities*.

19. Cross, "The Limits of Epistemic Communities."

20. Aronczyk, "Living the Brand"; "Understanding the Impact of the Transnational Promotional Class."

21. Edelman, *Politics of Misinformation*, 20.

22. For robust examinations of organizational coordination and influence by corporate actors in environmental politics, see Downie, "King Coal's Crown"; Barley, "Building an Institutional Field"; and secondarily, Hayden, Garner, and Hoffman, "Corporate,

Social and Political Networks of Koch Industries Inc. and TD Ameritrade Holding Corporation."

23. Fortun, "From Bhopal to the Informating of Environmentalism."
24. Fortun, "Biopolitics and the Informating of Environmentalism."
25. For a strong account of corporate power and its public relations in both the United States and the United Kingdom from the First World War through the twenty-first century, see Miller and Dinan, *A Century of Spin.*
26. Marchand, *Creating the Corporate Soul*; Tiffany, "Corporate Management of the 'External Environment.'"
27. Ida Tarbell, *The History of the Standard Oil Company*; Upton Sinclair, *Oil!*
28. Miller, *The Voice of Business.*
29. Berry, *Lobbying for the People*; Bob, *The Marketing of Rebellion*; Demetrious, *Public Relations, Activism, and Social Change*; Bosso, *Environment, Inc.*; Thomson and John, eds., *New Activism and the Corporate Response.* A full list of interviews conducted for this book appears in Appendix 1.
30. McCright and Dunlap, "Anti-Reflexivity"; McCright, "Anti-Reflexivity and Climate Change Skepticism in the U.S. General Public"; Dunlap and McCright, "Organized Climate Change Denial"; McCright and Dunlap, "The Politicization of Climate Change and Polarization in the American Public's Views of Global Warming, 2001–2010."
31. Pulver, "Making Sense of Corporate Environmentalism."
32. Edward Walker, *Grassroots for Hire*; Tim Wood, "Corporate Front Groups and the Making of a Petro-Public."
33. Of course, this same strategy attends journalistic coverage. See Pooley, *The Climate War.*
34. See Aronczyk, "Public Relations, Issue Management, and the Transformation of American Environmentalism, 1948–1992," for an example of PR work to establish such categorization around Rachel Carson's book, *Silent Spring.* During fieldwork at a conference for public relations counselors in the oil and gas industry in 2015, one PR firm presented slides detailing "the anatomy of an activist," compiling data on activist targets, issues, and actions to develop a profile of the category for industrial clients.
35. See Appendices 1 and 2 for a full list of interviews and fieldwork sites.
36. On the gendered hierarchy of the public relations profession, see Fitch, "The PR Girl," and Daymon and Demetrious, eds, *Gender and Public Relations*; on power in public relations, see Edwards, *Power, Diversity, and Public Relations.* On diversity and race in public relations, see Ford and Brown, "State of the PR Industry," and Munshi and Edwards, "Understanding 'Race' in/and Public Relations."
37. Clemens, *The People's Lobby*, 1.
38. Conley, "Environmentalism Contained."
39. Boltanski and Thévenot, *On Justification.*
40. On cultural and political framing of environmental movements and fields, see Brulle and Benford, "From Game Protection to Wildlife Management"; Lounsbury, Ventresca, and Hirsch, "Social Movements, Field Frames, and Industry Emergence." On disciplining discourses of environmental governance, see Bartley, "How

Foundations Shape Social Movements"; Broome and Quirk, "Governing the World at a Distance: The Practice of Global Benchmarking"; Brown, De Jong, and Lessidrenska, "The Rise of the Global Reporting Initiative: A Case of Institutional Entrepreneurship." On the strategic uses of political communication and public opinion management around the environment (among other public policy issues), see Manheim, *Strategy in Information and Influence Campaigns*; Bennett and Iyengar, "A New Era of Minimal Effects?"; Uldam, "Activism and the Online Mediation Opportunity Structure." On rhetorical and image strategies to narrate and visualize environmentalism, see DeLuca, *Image Politics*; Dunaway, *Seeing Green*; Schneider et al., *Under Pressure*; Matz and Renfrew, "Selling Fracking"; LeMenager, *Living Oil*; Gismondi and Davidson, "Imagining the Tar Sands 1880–1967 and Beyond." On technologies of environmental informating, modeling, mapping, and monitoring, see Gabrys, *Program Earth*; Fortun, "From Bhopal to the Informating of Environmentalism."

41. Warner, *Public and Counterpublics*, 72.
42. Mitchell, *Carbon Democracy*, 9.
43. Mitchell, *Carbon Democracy*, 3.

Chapter 1

1. Ewen, *PR!*, 50.
2. Lloyd, "The Story of a Great Monopoly." The business historian Thomas McCraw locates the origins of narratives documenting the adversary relationship between public and private spheres in the Progressive era, noting that Progressive history from 1901 to 1914 "recast the American experience as a continuous contest between public and private interests; that is to say, between right and wrong." McCraw, "Business & Government: The Origins of the Adversary Relationship," 40.
3. Schudson, *Discovering the News*.
4. Schudson discusses the simultaneous rise of news as entertainment, or "storytelling," and as "informational ideal" marked by "fairness, objectivity, and scrupulous dispassion." The major difference between the two genres lay in the social and political orientations of its readers as well as the professionalization of the industry of news. This is relevant to our account for a number of reasons: first, the valuation of "information" over "story" as a rationalized, fact-based endeavor was itself a moral project of elevating the news profession. The same trajectory can be seen here with both publicity and the idea of the environment. Second, the ideal of news as factual information reflects the elitism and conservatism of the brand of environmentalism (and publicity) that Americans have inherited. See Schudson, *Discovering the News*, 90.
5. We must include here Native Americans and slaves, who were treated as part of the "uncivilized nature" over which colonizers had mastery. Brulle, *Agency, Democracy, and Nature*, 117.
6. Buell, "Toxic Discourse."

7. Turner, "The Significance of the Frontier in American History." The original version of this essay was presented at the 1893 meeting of the American Historical Association in Chicago and published in the 1893 Annual Report of the American Historical Association. A longer version was subsequently published in Turner's essay collection, *The Frontier in American History*. It is fundamentally important to acknowledge that these environmental "origin" myths, of which Turner's is only one, relied on the suppression or erasure of the environment's original inhabitants. Indigenous peoples were consistently depicted as being part of the physical nature that needed to be taken in hand by the land's colonizers. It was by the elimination of Native Americans and the control of slave labor that nature could appear unpeopled and in need of protection.

8. Cronon, "The Trouble with Wilderness," 21. Cronon is referring as well to the historical fact of removal of the original inhabitants of the land.

9. The legacy of John Muir has been subject to rethinking in our time in light of his own racist treatment of Native Americans and African Americans. See, e.g., Fears and Mufson, "Liberal, Progressive—and Racist?"

10. Muir to Mrs. Ezra S. Carr, 7 October 1874; quoted in Muir, *Travels in Alaska*.

11. Johnson, *Remembered Yesterdays*, 284.

12. Nash, *Wilderness and the American Mind*, chapter 8; Johnson, *Remembered Yesterdays*.

13. Johnson, *Remembered Yesterdays*, 112.

14. Johnson, *Remembered Yesterdays*, 240.

15. Muir, "The Treasures of the Yosemite"; Muir, "Features of the Proposed Yosemite National Park."

16. Johnson, *Remembered Yesterdays*, 287–88.

17. Nash, *Wilderness and the American Mind*, 132–33.

18. Johnson, *Remembered Yesterdays*, 291.

19. Johnson, "A Plan to Save the Forests," 626.

20. Gifford Pinchot was also one of the opinion writers in this series. In hindsight, it seems that Pinchot was rather cautious in his approval of the vision, supporting a school of forestry "established at West Point or elsewhere." Johnson, "A Plan to Save the Forests," 630.

21. "Topics of the Time: The Need of a National Forest Commission," 635.

22. Nash, *Wilderness and the American Mind*, 136.

23. Hays, *Conservation and the Gospel of Efficiency*, 36–37.

24. Muir, "The American Forests"; Muir, "The National Parks and Forest Reservations."

25. Muir, "The Wild Parks and Forest Reservations of the West."

26. Nash, *Wilderness and the American Mind*, 139.

27. Morris, *Theodore Rex*, 230–31.

28. Johnson, *Remembered Yesterdays*.

29. Lippmann, *A Preface to Politics*, 1913.

30. Garey and Hott (dirs.), "The Wilderness Idea."

31. Pinchot to R. C. Melward, 20 May 1903, Office of Forest Reserves Correspondence; quoted in Pinkett, *Gifford Pinchot: Private and Public Forester*, 53.

32. Scott, *Seeing Like a State*.

33. Peters, "Democracy and American Mass Communication Theory."

34. Johnson, *Remembered Yesterdays*, 315; Fox, *John Muir and His Legacy*.

35. Pinkett, *Gifford Pinchot*, 3.

36. Scott, *Seeing Like a State*, 19.

37. Miller, *Gifford Pinchot and the Making of Modern Environmentalism*, 103.

38. Pinkett, *Gifford Pinchot*, 26. Frederick Jackson Turner delivered his speech, "The Significance of the Frontier in American History," at this same Chicago World's Columbian Exposition (as well as at the American Historical Association meeting in Chicago). Amid the hundreds of new technologies, products, and inventions (including the electric light bulb) were presentations of end of the frontier and the beginning of forestry.

39. Pinkett, *Gifford Pinchot*, 26; McGeary, *Gifford Pinchot*, 31.

40. "Mr. Vanderbilt's Forest," *Garden and Forest*, 7.313, 21 February 1894, 71. Cited in Pinkett, *Gifford Pinchot*, 26.

41. Pinkett, *Gifford Pinchot*, chapter 5.

42. Ponder, "Gifford Pinchot: Press Agent for Forestry," 28. For Pinchot's own perspective on the situation, see Pinchot, "Part IV: The President Makes the Issue," pp. 105–32 in *Breaking New Ground*.

43. Pinchot said the experience gave him "some inkling into how public opinion is credited or directed." Ponder, "Gifford Pinchot: Press Agent for Forestry," 28.

44. Pinkett, *Gifford Pinchot*, chapter 7.

45. Pinkett, *Gifford Pinchot*, 48.

46. Ponder, "Gifford Pinchot: Press Agent for Forestry," 28.

47. Pinkett, *Gifford Pinchot*, 48–50.

48. Pinkett, *Gifford Pinchot*, 53.

49. Poovey, *Genres of the Credit Economy*, 80.

50. Sheingate, *Building a Business of Politics*, 16.

51. Pinkett, *Gifford Pinchot*, 82.

52. Pinchot (Forest Service, US Department of Agriculture) to Hon. Charles F. Scott (Chairman, Committee on Agriculture, House of Representatives), on the matter of using the Forestry Service resources for publicity. Congressional Record: House (30 March 1908): 4138.

53. "We prepare the news—the valuable information that is news—in such shape that the newspapers will take it, not in any sense puffing our work; simply a definite statement of facts. The newspaper men come around and get that and print it. In that way we are getting before the people, with an utterly insignificant cost—two men do all this work, and they do not spend their whole time at it—material in an amount which would cost us thousands upon thousands of dollars every year to get out if we mailed it ourselves." Hearings before the Committee on Agriculture, Agricultural Appropriations Bill, 60th Cong., 1st sess. (1908): 276–77.

54. Pinkett, *Gifford Pinchot*, 84.

55. Ponder, "Gifford Pinchot: Press Agent for Forestry," 35.

56. Roosevelt, *Theodore Roosevelt: An Autobiography*, chapter 11.

57. Pinkett, *Gifford Pinchot*, 53.

58. Dennehy, "First Forester: The Enduring Conservation Legacy of Gifford Pinchot."

59. Miller, *Gifford Pinchot*, 196. That same year, Pinchot founded the Society of American Foresters, along with the first *Journal of Forestry*, to establish "professional standards in forestry." Pinkett, *Gifford Pinchot*, 87. Its original members, mainly Yale classmates, would also find their way into roles in the Forest Service. See Gonzalez, *Corporate Power and the Environment*, for an expanded discussion of the professionalization of the environmental policy network.

60. See Gonzalez, "The Conservation Policy Network, 1890–1910."

61. Pinkett, *Gifford Pinchot*, 36–37; 41.

62. See Ross, "From Practical Woodsman to Professional Forester."

63. Gonzalez, "The Conservation Policy Network," 277.

64. Miller, *Gifford Pinchot*, 220.

65. Ponder, "Gifford Pinchot: Press Agent for Forestry," 28.

66. Pinkett, *Gifford Pinchot*, 69.

67. Miller, *Gifford Pinchot*, 226. In his autobiography, Theodore Roosevelt had high praise for Woodruff: "The idea that the Executive is the steward of the public welfare was first formulated and given practical effect in the Forest Service by its law officer, George Woodruff." Quoted in Pinkett, *Gifford Pinchot*, 69.

68. Miller, *Gifford Pinchot*, 227. The multiple organizational structures provided the appearance of broad and varied support for his brand of conservationism; and also broadened the concept of the environment, connecting water power to forestry. This would matter considerably during the battle over water use in the Hetch-Hetchy Valley.

69. Yates, "Creating Organizational Memory."

70. Yates, "Creating Organizational Memory." Pinchot may have encountered vertical filing systems for the first time on display at the Chicago World's Fair in 1893, where he began his publicity for the forests. See Pinkett, "The Forest Service: Trail Blazer in Recordkeeping Methods," 421–424.

71. Ponder, "Progressive Drive to Shape Public Opinion," 97.

72. Miller, *Gifford Pinchot*, 158–59.

73. Miller, *Gifford Pinchot*, 157–58; Pinchot, "The Use of the National Forests." But see Steen, *Forest Service: A History*.

74. Miller, *Gifford Pinchot*, 228.

75. Righter, *The Battle over Hetch-Hetchy*, 215; Cronon, "The Trouble with Wilderness," 9. There is cruel irony in calling this story any of these things. The story obliterates the primary loss by Native Americans of a 200-year connection to the land, not only in material terms but also as a sacred homeland. As anthropologist Bruce Pierini writes, "The loss of homelands at Hetch-Hetchy is, at the most profound level, a loss of a centuries-old way of life sustained by an empirically based yet mystical worldview." Pierini, "How Did the Hetch-Hetchy Project Impact Native Americans?"

76. Sewell, *Logics of History*, 236.

77. US Congress, "Chapter 372: An Act Relating to Rights of Way," 56th Cong., 2nd sess. (15 February 1901): 791.

78. Nash, *Wilderness and the American Mind*, 161.

79. "Begin Fight to Save the Yosemite Park," 8.

80. Clemens, *The People's Lobby*, 28.
81. Johnson, *Remembered Yesterdays*, 309.
82. See also Oravec, "Conservationism versus Preservationism: The 'Public Interest' in the Hetch-Hetchy Controversy," who notes this same discursive tactic in Pinchot's testimony in the 1912 hearings.
83. Johnson, "A High Price to Pay for Water"; and Johnson, *Remembered Yesterdays*, 311.
84. Nash, *Wilderness and the American Mind*, 169.
85. Quoted in Nash, *Wilderness and the American Mind*, 174. This was clearly a politically motivated act. Kent was an avid hunter and a clear proponent of Roosevelt's (that is, Pinchot's) views on conservation. But he shared with his friend Muir a love of unspoiled wilderness, helping to establish the Muir Woods National Monument in 1908.
86. Johnson, "A High Price to Pay for Water," 663.
87. Oravec, "Conservationism versus Preservationism," 453.

Chapter 2

1. Mitchell, *Carbon Democracy*. In *The Public and Its Problems* (1927), Charles Dewey makes this observation: "Invent the railway, the telegraph, mass manufacture and concentration of population in urban centers, and some form of democratic government is, humanly speaking, inevitable" (110). But while the infrastructure permits the possibilities of democratic politics, Dewey cautions, its corollary—democratic publics—must come out of community and association.
2. Tedlow, *Keeping the Corporate Image*.
3. Mitchell, *Carbon Democracy*, 19–20; 26–27.
4. In addition, the shift from coal to oil de-localized its work force. Migratory and temporary laborers could not organize the way local laborers had. Mitchell, *Carbon Democracy*, chapter 1. See also Bowker, *Science on the Run*; Wylie, Shapiro, and Liboiron, "Making and Doing Politics through Grassroots Scientific Research on the Energy and Petrochemical Industries."
5. Freudenberg and Alario, "Weapons of Mass Distraction."
6. See, e.g., Tiffany, "Corporate Management of the 'External Environment.'" Today, this idea is better known as a "social license" for companies to operate. Aronczyk, "Understanding the Impact of the Transnational Promotional Class."
7. Bernays and Ivy Lee both used the term "counsel," with Bernays credited for coining it in 1913. Hiebert, *Courtier to the Crowd*, 87.
8. A good primer for understanding the industrial context that gave rise to the idea of industrial democracy is the 2016 documentary film *The Mine Wars*, directed by Randall MacLowry.
9. In this sense we can also think of the era's public relations as anticipating some of the observations of media infrastructural studies, such as the "politics of infrastructural invisibility" by which material infrastructures are disguised as part of the natural environment, or the ways that commercial data centers are retooled as sources of climate

action, which will be the focus of chapter 7. See Parks, "Around the Antenna Tree"; Brodie, "Climate Extraction and Supply Chains of Data"; Bowker, Baker, Miller, and Ribes, "Toward Information Infrastructure Studies: Ways of Knowing in a Networked Environment."

10. These hagiographers were both PR proponents (by which we mean here industry sympathizers and conservatives) and PR men themselves, whose own multiple promotional publications about their clients and their tactics of persuasion served as sources of information for these biographies.

11. Hiebert, *Courtier to the Crowd*, 8–9.

12. Habermas, *The Structural Transformation of the Public Sphere*, 194. Habermas conflates advertising and public relations, which leads him to portray PR as uniquely about the promotion of private (mainly commercial) interests for political purposes. It also leads him to minimize the effects of public relations by treating it as a system of mediated messaging rather than as a structural phenomenon.

13. Habermas, *Structural Transformation*, 192–93; 201. There is actually a third source of writings that constitute Lee's historical legacy: Lee's own extensive documentation of his publicity work. In his lifetime Lee wrote hundreds of speeches, pamphlets, and articles, many of which were collected and reprinted in book form. Some covered topics germane to his clients, such as railway histories and tracts about their economic potential; but many of his writings dealt with the topic of publicity itself, creating a benchmark against which other emerging PR practitioners had to define themselves.

14. Hiebert, *Courtier to the Crowd*, 151.

15. Hiebert, *Courtier to the Crowd*, 149; see also Olasky, *Corporate Public Relations*.

16. Hiebert, *Courtier to the Crowd*, Appendix C: 338–42.

17. Mitchell, *Carbon Democracy*, 19.

18. Hiebert, *Courtier to the Crowd*, 91.

19. One campaign for the railroads involved promoting the scenic, industrial, and agricultural benefits in California to help encourage continued industrial growth in the region. Hiebert, *Courtier to the Crowd*, 59.

20. Warner, *Publics and Counterpublics*.

21. US Congress, 63rd Cong., 2nd sess., *Congressional Record* 51 (5 May 1914): 7729.

22. US Congress, 63rd Cong., 2nd sess., *Congressional Record* 51 (5 May 1914): 7729–30. See also Hiebert, *Courtier to the Crowd*, 65.

23. US Congress, 63rd Cong., 2nd sess., *Congressional Record* 51 (5 May 1914): 7818.

24. Cronon, *Nature's Metropolis*.

25. US Congress, 63rd Cong., 2nd sess., *Congressional Record* 51 (5 May 1914): 7738.

26. "Maintenance of a Lobby to Influence Legislation" (Hearings before a Subcommittee on the Judiciary, US Senate), 63rd Cong., 1st sess. (25 June 1913): 1665.

27. US Congress, 63rd Cong., 2nd sess., *Congressional Record* 51 (5 May 1914): 7749.

28. US Congress, 63rd Cong., 2nd sess., *Congressional Record* 51 (5 May 1914): 7729.

29. Lee, "Enemies of Publicity."

30. Hallahan, "Ivy Lee and the Rockefellers' Response to the 1913–1914 Colorado Coal Strike," 266.

31. Hallahan, "Ivy Lee," 269; 270n6.

32. Hallahan, "Ivy Lee," 271–72. Hallahan also describes how the union copied this tactic, making its own set of bulletins that looked exactly like Lee's but gave the union position.

33. Public relations historian Kirk Hallahan called Mackenzie King "Rockefeller's 'other' public relations counselor in Colorado." Hallahan, "W. L. Mackenzie King," 401.

34. Hiebert, *Courtier to the Crowd*, 98. See also the documentary film, *The Image Makers*, directed by David Grubin.

35. Marchand, *Creating the Corporate Soul*, 16.

36. Hallahan, "Ivy Lee," 279.

37. Hiebert, *Courtier to the Crowd*, 102.

38. Hallahan, "Ivy Lee," 278.

39. United States Department of the Interior National Park Service, "Ludlow Tent Colony Site," 52. See also Andrews, *Road to Ludlow*; Gitelman, *Legacy of the Ludlow Massacre*.

40. One particularly blatant example is found in the *New York Times* on 15 September 1915, titled "Rockefeller Plies Pick in Coal Mine; Dons Overalls and Jumper and Makes First-Hand Observations of Colorado Conditions; Calls Men His Partners; Tells Them Their Interests Are Similar; Questions Coal Diggers about Wages and Work." A popular photograph of John D. Rockefeller and William Lyon Mackenzie King touring the CF&I mine in Valdez, Colorado, was also featured in the *New York Times* on 5 October 1915.

41. Domhoff, "The Rise and Fall of Labor Unions."

42. Barenberg, "Democracy and Domination in the Law of Workplace Cooperation," 806–7.

43. US Congress, "A Resolution to Investigate Violations of Free Speech and Assembly" (Hearings before a Subcommittee of the Committee on Education and Labor, US Senate), 76th Cong., 1st sess., Part 37: Supplementary Exhibits (16 January 1939): 15782. Emphasis added.

44. Given the interconnectedness of coal, steel, rail, and oil in terms of industrial production as well as in ownership and intra-sector coordination, these industrial sectors were all operating along similar lines. The focus on the steel industry here is an analytical separation, not a functional one.

45. Warren, *The American Steel Industry*.

46. Mumford, "This Land of Opportunity."

47. Quoted in Spillman, *Solidarity in Strategy*, 47. See also Bradley, *Role of Trade Associations and Professional Business Societies in America*; and Roy and Parker-Gwin, "How Many Logics of Collective Action?"

48. *National Industrial Recovery Act*, HR 5755, 73rd Cong., 1st sess., 1933.

49. Originally three addresses delivered between 1916 and 1925 to various public audiences, as well as the transcript of the question and answer period following the speeches.

50. Lee, *Publicity*, 19–20.

51. Lee, *Publicity*, 20.

52. Hill, *The Making of a Public Relations Man*, 61.

53. "A Resolution to Investigate Violations of the Right of Free Speech and Assembly and Interference with the Right of Labor to Organize and Bargain Collectively" (Hearings before a Subcommittee of the Committee on Education and Labor, US Senate), 76th Cong., 1st sess., Part 40: Supplemental Exhibits: Hill and Knowlton, Public Relations Counsel (16 January 1939): 15560.
54. "Golden Interview with John W. Hill" (Part I), 17.
55. National Labor Relations Board, "1935 Passage of the Wagner Act."
56. "Golden Interview with John W. Hill" (Part II), 2.
57. US Congress, "A Resolution to Investigate Violations" (Part 40), 15553.
58. US Congress, "A Resolution to Investigate Violations" (Part 38), 15546.
59. Mumford, "This Land of Opportunity."
60. White, *The Last Great Strike*, 192.
61. "What Is the N.A.M.?," 20.
62. Blumenthal, "Anti-Union Publicity in the Johnstown 'Little Steel' Strike of 1937," 677.
63. US Congress, "A Resolution to Investigate Violations," 204. Cutlip (*The Unseen Power*, 467) writes that "in this period the NAM (National Association of Manufacturers) became a client of Hill & Knowlton. The NAM and AISI (American Iron and Steel Institute) worked hand-in-glove in combating organization of steel by the SWOC (Steel Workers Organizing Committee)."
64. On the use of civic rationales to justify business dealings, see Boltanski and Thévenot, *On Justification*.
65. Citizens' committees were already being formed around industry labor issues prior to the Little Steel Strike. For instance, a strike at the Goodyear Tire & Rubber Co. in Ohio in 1936 was met with a letter-writing campaign by a citizens' committee, calling for an end to the strike to "maintain industrial peace and progress in this community." See US Congress, "A Resolution to Investigate Violations" (Part 40), 15605.
66. Blumenthal, "Anti-union Publicity."
67. The booklet, *The Men Who Make Steel*, was from Sokolsky's pen. US Congress, "A Resolution to Investigate Violations" (Part 40), 204.
68. "Self-Evident Subtlety."
69. "Golden Interview with John W. Hill."
70. See, e.g., Hill, "What We Learned from the Steel Negotiations."
71. As Karen Miller writes in *The Voice of Business*, "The early history of the Tobacco Industry Research Committee (TIRC) and the Tobacco Institute is indivisible from the history of Hill and Knowlton" (131). The TIRC's executive director was on the payroll of Hill & Knowlton, and the contacts listed in promotional materials for the tobacco associations were Hill & Knowlton staff.
72. Spillman, *Solidarity in Strategy*, 297.

Chapter 3

1. Sewell, "Historical Events as Transformations of Structures."
2. Prior to around 1966, environmentalists were mainly called "conservationists" (this latter title reflecting the triumph of Gifford Pinchot's rationalized perspective on nature as resource to be managed rather than John Muir's more communal perspective of preservation). It was through the writings of ecologists such as Barry Commoner and Paul and Anne Ehrlich as well as Rachel Carson that a "reform" environmentalism surfaced—"the insight that humanity is part of the earth's ecosystems and thus human health is linked to the condition of the natural environment." Carmichael, Jenkins, and Brulle, "Building Environmentalism," 452.
3. Murphy, *What a Book Can Do*. For a strong audiovisual account of the attempt by industrial public relations to discredit Carson and her book, see the documentary film *Rachel Carson*, directed by Michelle Ferrari.
4. Marchand, *Creating the Corporate Soul*, chapter 6.
5. Andrew Hurley, *Environmental Inequalities*, quoted in Conley, *Environmentalism Contained*, 44.
6. Conley, "Environmentalism Contained." See also Jasanoff, "Procedural Choices in Regulatory Science."
7. Conley, "Environmentalism Contained," 12–13. See also Sellers, *Hazards of the Job*, 1997; Hounshell and Smith, *Science and Corporate Strategy*.
8. Jasanoff, *The Fifth Branch*; Bocking, *Nature's Experts*.
9. The AISI had initiated a research program on air pollution "after Allegheny County in Pennsylvania, home to Pittsburgh's steel industry, passed an ordinance in 1949 mandating research by local steel firms." The American Petroleum Institute's Smoke and Fumes Committee was sponsoring multiple research projects by the mid-1950s. Conley, *Environmentalism Contained*, 22–23.
10. Though they paid a $4.6 million settlement, the company denied responsibility, blaming a "freak weather condition" and a broad set of smog producers from "homes, railroads, steamboats, and the exhaust from automobiles." See "Steel Company Pays $235,000 to Settle $4,643,000 in Donora Smog Death Suits."
11. Ross and Amter, *The Polluters*, 147. For the industry perspective, see Best, "A Rational Approach to Air Pollution Legislation."
12. Hull, "Accomplishments in Air Pollution Control by the Chemical Industry." Retired US Army General John J. Hull was president of MCA from 1955 to 1961.
13. Quoted in Conley, "Environmentalism Contained," 55–56; see also Ross and Amter, *The Polluters*, 147–48.
14. John E. Hull, "Accomplishments in Air Pollution Control by the Chemical Industry," 64.
15. Conley, "Environmentalism Contained," 56–57.
16. Harrison, "Environmental Health Committee Meeting."
17. Grunig, "Review of Research on Environmental Public Relations." *Public Opinion Quarterly* polls included those by Simon, "Public Attitudes Toward Population and Pollution," and Erskine, "The Polls: Pollution and Its Costs," as well as Erskine, "The

Polls: Pollution and Industry." See also Tichenor et al., "Environment and Public Opinion."

18. On the emergence of institutional public opinion, see Sudman and Bradburn, "The Organizational Growth of Public Opinion Research in the United States." On the use of public opinion polling in politics, see Johnson, *Democracy for Hire*. On the relationship of surveys and polls to the making of the American public, see Igo, *The Averaged American*.

19. Harrison was promoted the following year to vice-president. The company is today called Freeport-McMoRan.

20. Stuart Kirsch, *Mining Capitalism*. The Manufacturing Chemists' Association (MCA) was renamed the Chemical Manufacturers Association (CMA) in 1978. Along with the name change came a change in scale and scope of the trade association. The CMA hired a new CEO, Bob Roland, with decades of experience in Washington, DC; increased dues revenue from $4 million to $7.5 million; and doubled its staff size. In 2000 the name changed again, to the American Chemistry Council.

21. "D.C. Agency Created First Client."

22. E. Bruce Harrison's spouse, Patricia Harrison, was co-founder of Harrison & Associates, and was vice-president of the firm.

23. Lerbinger, "A Long View of the Environment."

24. In addition to its testimony on the Clean Air Act, NEDA also conducted lobbying efforts on behalf of the Poage-Wampler bill (pesticide control) in 1975. Introduced by Representative W. R. Poage (D. Tex.) and Representative. William Wampler (R. Va.), members of the House Agriculture Committee, the measure was intended to allow the Agriculture Department to retain authority over decisions regarding pesticide control instead of the Environmental Protection Agency. See Anderson and Whitten, "The Washington Merry-Go-Round." NEDA also appeared in the hearings before the Subcommittee on Water Resources of the committee on public works and transportation in 1977 amid efforts to amend the Federal Water Pollution and Control Act.

25. "Statement of Thomas A. Young—Clean Air Act Oversight," 93rd Cong., 1st sess. (1973), 1045.

26. Anderson and Whitten, "Washington Merry-Go-Round."

27. As Thomas A. Young, president of NEDA, explained during his testimony: "As to this matter before the Subcommittee, we espouse as all men must, the objectives of the Clean Air Act—and those other objectives of full employment and economic growth essential to the general welfare of all Americans. In these and other statutory matters affecting the human environment, it is our insistent view, however, that each be pursued in a manner compatible with the attainment of the others." "Statement of Thomas A. Young," 1045. See also Robert Kerr, *The Rights of Corporate Speech*, 53, which describes how Mobil Oil also espoused this rhetoric of "balance."

28. Lerbinger, "A Long View of the Environment."

29. Awad, "Environment: A Continuing Arena." Joseph F. Awad was chairman of the Committee for the Environment of the Public Relations Society of America and General Director of Public Relations for the Reynolds Minerals Company in Richmond, Virginia.

30. Thompson, "Communicators and Their Environmental Problems," 34.
31. Galler and Littin, "Economic Impact: Perspectives for Corporate Decision-Making."
32. Lerbinger, "A Long View of the Environment," 20–21.
33. Buell, "Toxic Discourse," 650.
34. See chapter 6 of Davidson and Gismondi, *Challenging Legitimacy*, for a discussion of the same discourse of scarcity as justification for oil development in the twenty-first century.
35. Mitchell, *Carbon Democracy*, chapter 7.
36. Mitchell, *Carbon Democracy*, 191.
37. "PRSA–White House Conference on Energy," 6. This editorial appeared in the July 1974 issue of *Public Relations Journal*, which was devoted to the theme of "Communicating the Energy Crisis."
38. Mobil Oil, "Evolution of Mobil's Public Affairs Programs 1970–81," I-A/5.
39. Useem and Zald, "From Pressure Group to Social Movement," 151.
40. Shants, "Countering the Anti-Nuclear Activists." David Sicilia, in "The Corporation Under Siege: Social Movements, Regulation, Public Relations, and Tort Law Since the Second World War," has similarly shown the correspondence of campaign tactics across three contentious industries: chemical, tobacco, and nuclear energy, in their efforts to counter public and political pressure.
41. We are inspired here by Douglas Rogers's notion of "corporate social technologies," with their dual focus on sociability and materiality. See Rogers, "The Materiality of the Corporation."
42. Conference speakers included former EPA administrator William Ruckelshaus, Petr Beckmann of the University of Colorado, Robert White-Stevens of Rutgers University, Irwin Tucker of the University of Louisville, Senator Jennings Randolph (D. W.Va.), Representative John Rhodes (R. Ariz.), Michael Moskow, director of the Council on Wage and Price Stability, and Alvin Alm, assistant administrator for planning and management at the EPA.
43. "Maintaining an Environmental Balance," 418.
44. There had been, as early as 1973, a PRSA Task Force on Environment. Harrison's group was the one that introduced "energy" into the equation. See "Environment: A Continuing Arena," 2; and Harrison, "Environment Energy: Public Relations at Large."
45. Trade journals include *Professional Remodeling: The Monthly Management and Marketing Magazine for Today's Improvement and Expansion Contractor*, and *Hydrocarbon Processing*.
46. "D.C. Agency Created First Client."
47. "Kenneth Bousquet Dies, Former Senate Counsel," B7.
48. See Sicilia, "The Corporation under Siege."
49. Quarles's keynote address to the 1981 Annual Convention of the Air Pollution Control Association was titled "Maturing Environmentalism."
50. Harrison would later call environmental regulatory or legislative issues "greening issues." Personal communication, Harrison to Aronczyk, 2018.
51. William Haum, chair and vice-president of General Mills, helped found and chair NEDA-CWP.

52. Examples include Federal Water Pollution Control Act, HR 3199, 95th Cong., 1st sess. (1–4 March 1977); NEDA Clean Air Act Project, 1980–81 Plan; Quarles, "The Clean Air Amendments"; Quarles, "A Thicket of Environmental Laws"; Quarles, "EMB: Congress at Its Worst"; "Two Views: National Environmental Development Association"; "Cleaning Up the Clean Air Act: National Clean Air Coalition."
53. Conley, "Environmentalism Contained," 2.
54. Several of the issue papers were assembled into a workbook, *Clean Air Act & Industrial Growth*, and distributed broadly to members of Congress.
55. "Two Views: National Environmental Development Association."
56. Quarles, "The Clean Air Amendments."
57. Harrison, "Is 'No Growth' Really Ahead?" Emphasis in original. See also Harrison, "Clean Air Act," and "EPA Reaches Out."
58. "Grassroots Involvement: Key to Issue Management."
59. "Grassroots Public Relations."
60. Mitchell, *Carbon Democracy*, 176.

Chapter 4

1. Not including affiliates. See "Golden Interview with John Hill," 304. See also "John W. Hill, 86, Dies; Led Hill & Knowlton," 43.
2. It was not just that companies saw their functions as being aligned with the public interest; it was that they saw no conflict between private and public interests. See, e.g., Aronczyk, *Branding the Nation*, 42–3, on how the dictum, "What's good for General Motors is good for America," emerged as a justification for the GM president's nomination to U.S. Secretary of Defense under President Eisenhower.
3. Punctuated by a massive oil spill in Santa Barbara, California, in 1969, and four additional spills in the next several months. See Hoffman, *Heresy to Dogma*, 56.
4. Vogel, *Fluctuating Fortunes*, 65.
5. Habermas, *Structural Transformation*, chapter 6.
6. Vogel, *Fluctuating Fortunes*; Walker, "Legitimating the Corporation through Public Participation"; Mizruchi, *The Structure of Corporate Political Action*; Schuler, "Corporate Political Action: Rethinking the Economic and Institutional Influences"; Kay and Tierney, *Organized Interests and American Democracy*; Laumann and Knoke, *The Organizational State*.
7. Barley, "Building an Institutional Field."
8. Powell's Memorandum warned about the social and political threat to the American free enterprise system, arguing that corporations and industry must wrest control of the economy from leftist inclinations: "Conservatives must capture public opinion by exerting influence over the institutions that shape it: academia, media, church, courts." Powell, "Attack on the Free Enterprise System." Powell, who was in 1971 a corporate lawyer, would go on to become a Supreme Court justice, with considerable influence over such institutions. In 1973, top US Steel public relations counselor

William G. Whyte spoke before the PRSA. As part of a task force appointed by the president of the US Chamber of Commerce to coordinate actions around the Powell memo, he and other PRSA members directed the preparation of an information kit for distribution to local Chambers of Commerce (other kits were prepared for state chambers and for trade and professional associations). Called the Interpreting Business Kit, it contained guidelines for these organizations to promote the values of private enterprise to as many segments of society as possible: public relations advisors, as the managers of the image of business, were to play a central role: "We are living in a different world, one that makes the role of the public relations official ever more important. . . . [N]o place in the Nation does the slipping image of business come home to roost any more than it does in Washington, DC. There—image, power, and influence are pretty closely related. And when one slips, so does the other." Whyte, "Remarks before Public Relations Society of America."

9. On corporate grassroots strategies, see Walker, *Grassroots for Hire*; on cooperative oligopolies, see Munkirs & Sturgeon, "Oligopolistic Cooperation"; on interlocking directorates, see Hayden et al., "Corporate, Social and Political Networks."

10. In this sense, we may see PR actors' positionings of authority as engaging in the same kinds of politics underlying the legitimacy contests of science advisory committees. See Jasanoff, *The Fifth Branch*.

11. Edwards and Hodges, eds., *Public Relations, Society and Culture*. 3. This is different from the Habermasian understanding that PR is a kind of staging: "The consensus-concerning behavior required by the public interest, or so it seems, actually has certain features of a staged 'public opinion.' . . . The resulting consensus, of course, does not seriously have much in common with the final unanimity wrought by a time-consuming process of mutual enlightenment, for the 'general interest' on the basis of which alone a rational agreement between publicly competing opinions could freely be reached has disappeared precisely to the extent that the publicist self-presentations of privileged public interests have adopted it for themselves." Habermas, *Structural Transformation*, 193–95.

12. Boltanski and Thévenot, *On Justification*, 278.

13. Boltanski and Thévenot, *On Justification*, 281.

14. Dunlap, *DDT: Scientists, Citizens, and Public Policy*, 3. For an example of cinematic footage, see *The Story of DDT*.

15. Berry, *Lobbying for the People*; Clemens, *The People's Lobby*; Vogel, "The Public Interest Movement and the American Reform Tradition."

16. See David Vogel, "The Public Interest Movement."

17. Bosso, *Environment, Inc.*, 42.

18. Dunlap, *DDT*.

19. Carroll, "Participatory Technology," 649.

20. Carroll, "Participatory Technology," 649.

21. The two cases were Scenic Hudson Preservation Council v. the Federal Power Commission (1965), also known as the Storm King case; and Sierra Club v. Morton (1972), whose opinion was drafted by the Supreme Court based on the Storm King case. Lambert, "Scenic Hudson and Storm King."

22. Carroll, "Participatory Technology," 650.

23. See Melnick, *Regulation and the Courts.*

24. "Through their mere choice of words, self-described defenders of the 'public' interest implicitly condemned the 'private' sector for its inability to protect consumers, citizens, and the environment. And no one person typified that animus more than a young lawyer named Ralph Nader." Waterhouse, *Lobbying America,* 38.

25. Whiteside, "Profiles: A Countervailing Force—I," 84.

26. Drew, "A Reporter at Large: Conversation with a Citizen," 39.

27. Björk, "Emergence of Popular Participation in World Politics."

28. Lesly, "Survival in an Age of Activism," 8.

29. Moore, "Environment: A New PR Crisis," 7.

30. Brandt, "Wanted: Environmentalists," 19.

31. Hill & Knowlton, "Slings and Arrows," 2.

32. Hill & Knowlton, "Slings and Arrows," 4.

33. Hill & Knowlton, "Slings and Arrows," 31–32.

34. Wessel, *Science and Conscience,* 34.

35. Wessel, *Science and Conscience,* 28.

36. Wessel, *Science and Conscience,* 201; Wessel, *Rule of Reason,* 21; 202.

37. Parisi, "Book Brings the Rule of Reason to Corporation-Public Clashes."

38. Wessel, *Rule of Reason,* xi.

39. Sethi, "Corporate Political Activism," 40.

40. "New Ways to Lobby a Recalcitrant Congress," 148. See also Freed, "Melding PR and Lobbying Impact."

41. As Donald Colen, vice-president and director of public affairs of New York Citibank claimed: "In public relations now, all roads lead to the Hill" (quoted in Harrison, "Washington Focus"). This alliance between lobbyists and PR would shift again in the late 1980s in the aftermath of news investigations into "honoraria" paid to congresspeople. See Jackson, "Easy Money"; Kenworthy, "Courting the Key Committees."

42. "Juice: The Future of Power and Influence in Washington." See also Moore, "Have Smarts, Will Travel."

43. Wittenberg and Wittenberg, *How to Win in Washington.*

44. "New Ways to Lobby a Recalcitrant Congress," 148. See also Jones and Chase, "Managing Public Policy Issues," 9: "In the world of today, the diverse activities we call government and public relations, lobbying and issue advertising, must all be part of an integrated management strategy."

45. Harrison, "Washington Focus."

46. Swetonic, "Death of the Asbestos Industry," 9.

47. Similar initiatives took place around the same time within the dispute resolution forums at Harvard Negotiation Project, an initiative piloted in 1979 that led to the subsequent publication of *Getting to Yes,* by project leaders Roger Fisher and William Ury.

48. Dunlap, *DDT,* 235.

49. Dunlap, *DDT;* Conley, "Environmentalism Contained."

50. Wessel, *Science and Conscience*, 145.
51. Wessel, *Science and Conscience*, 142.
52. Wessel, *Science and Conscience*, 155.
53. Wessel, *Science and Conscience*, 157.
54. Wessel, *Science and Conscience*, 158.
55. Rich and Jacobson, "Alternative Dispute Resolution," 30.
56. Rich and Jacobson, "Alternative Dispute Resolution," 30–31.
57. Rich and Jacobson, "Alternative Dispute Resolution," 32.
58. Schudson, *Rise of the Right to Know*, 1.
59. Schudson, *Rise of the Right to Know*, 181, 185–86.
60. LeMenager, *Living Oil*, 45.
61. LeMenager, *Living Oil*, 45. See also the documentary film, *How to Change the World*, about Greenpeace's innovative adoptions of media and high-profile, dramatic events to gain public attention. Greenpeace's legacy is evident in more recent climate awareness campaigns by organizations such as Extinction Rebellion and popular movements such as school climate strikes. Its legacy was also apparent in some of our interviews with industrial public relations actors. One derisively characterized climate activists who used such publicity tactics as "these people who seem to be sensational opportunists that are trying to play upon the emotions without any real fact behind their arguments, because they just want to buy a boat"—in reference to Greenpeace's origins in 1971, when a group of activists sailed from Vancouver to Amchitka Island in Alaska in a fishing boat to protest President Nixon's nuclear weapons tests.
62. Brown and Waltzer, "Buying National Ink."
63. Brown and Waltzer, "Every Thursday," 25.
64. Schmertz and Novak, *Goodbye to the Low Profile*, 139.
65. Schmertz and Novak, *Goodbye to the Low Profile*, 20. For an overview of Schmertz's approach to public relations, see St. John III, "The 'Creative Confrontation' of Herbert Schmertz."
66. Sethi and Schmertz, "Industry Fights Back," 20.
67. Mobil Oil, "Evolution of Mobil's Public Affairs Programs," I-C/5.
68. Schmertz and Novak, *Goodbye to the Low Profile*, 145.
69. LeMenager, *Living Oil*, 146.
70. Brown and Waltzer, "Every Thursday," 200–201. Mobil continued to place advertorials after the year 2000, though with less frequency. Geoffrey Supran and Naomi Oreskes count Mobil's more recent advertorials (1989–2004) as part of a large-scale campaign by the company to sow doubt around climate science. See Supran and Oreskes, "Assessing ExxonMobil's Climate Change Communications."
71. Mobil Oil, "Evolution of Mobil's Public Affairs Programs," I-C/9.
72. Mobil Oil, "Evolution of Mobil's Public Affairs Programs," I-C/10.
73. Mobil Oil, "Evolution of Mobil's Public Affairs Programs," I-C/10; II-B/12.
74. Schmertz and Novak, *Goodbye to the Low Profile*, 210.
75. Mobil Oil, "Evolution of Mobil's Public Affairs Programs," 216.
76. Mobil Oil, "Evolution of Mobil's Public Affairs Programs," 221–30.
77. Jarvik, "PBS and the Politics of Quality," 265.

78. Kerr, *The Rights of Corporate Speech*, 2.

79. Chase, *Issue Management*, 6-7; Sonnenfeld, *Corporate Views*; David Rockefeller, "Free Trade in Ideas," *Chief Executive Magazine*; Aronczyk, "Public Relations, Issue Management, and the Transformation of American Environmentalism."

80. Harrison, "Green Communication."

81. Sethi, "Corporate Political Activism," 38.

82. Sethi, "Corporate Political Activism," 34. See also Sethi, "Serving the Public Interest."

83. Cohen, "Business Lobby," 1050.

84. McFarland, *Cooperative Pluralism*.

85. Vietor, *Environmental Politics and the Coal Coalition*.

86. The late 1960s and the 1970s were also a period in which coal miners took part in thousands of wildcat strikes. See, e.g., Turl, "The Miners' Strike of 1977-78."

87. National Coal Policy Project, Hearing before the Subcommittee on Energy and Power, 95th Cong., 2nd sess. (10 April 1978): 2-3.

88. Moyer, "Where We Agree," 971.

89. Quoted in Hoffman, *Heresy to Dogma*, 93.

90. "National Coal Policy Project a Mixed Success," 8.

91. Harrison, "Rule of Reason," 1.

92. Bosso, *Environment, Inc.*, 130.

93. Buchholz et al., *Managing Environmental Issues*, vii-xiii.

94. "We Can Work with You," 7.

95. "We Can Work with You," 7.

96. Libbey, "Conservation and the Corporation." For more on the "smooth operatives" of the Nature Conservancy, see Wood, "Business-suited Saviors of Nation's Vanishing Wilds."

97. "Union Camp, Georgia Pacific, and Dravo Donate Key Natural Areas," 1.

98. "We Can Work with You."

99. "Environmental Partnerships Help Business Find Effective Solutions."

Chapter 5

1. "Coordination with the United Nations System."

2. The Brundtland Report, formally titled *Our Common Future* (1987), laid out principles for sustainable development that united concerns of northern and southern countries. It famously defined sustainable development as "development that meets the needs of the present without compromising the ability of future generations to meet their own needs." It was prepared by the World Commission on Environment and Development, chaired by Gro Harlem Brundtland, former prime minister of Norway.

3. See Bernstein, *The Compromise of Liberal Environmentalism*, for a detailed examination of UNCHE and its impacts on the path formation of sustainable development for UNCED.

4. Bernstein, *The Compromise of Liberal Environmentalism*, 48.

5. Sklair, "The Transnational Capitalist Class and the Discourse of Globalization"; Levy, "Environmental Management as Political Sustainability"; Bernstein, *Compromise*.

6. Bernstein notes that some studies identify the same features but refer to it as "ecological modernization" or even just "sustainable development." *Compromise*, 7.

7. Pattberg, *Private Institutions and Global Governance*.

8. Pallemaerts, "International Environmental Law from Stockholm to Rio," quoted in Bernstein, *Compromise*, 50.

9. Note that the concept and term "sustainability" did not emerge with the UNCED. Already by the mid-1970s, the UN Environment Programme, led by Maurice Strong, was advocating sustainability (if not sustainable *development* as a specific term), and this notion contained within it a commitment to economic growth along with improvement of the living conditions in the developing world. Bernstein, *Compromise of Liberal Environmentalism*, 56.

10. There was non-American public relations action around environmental problems as well, primarily in Britain but also in Australia (where public relations emerged especially to address the activities of the mining industry; see Kirsch, *Mining Capitalism*). But by and large, public relations followed the patterns of colonial or quasi-colonial expansion common to most promotional industries, as corporate outposts of major media and marketing conglomerates such as Omnicom, WPP, or Edelman set up shop across international territories. These patterns are examined in Miller and Dinan, *A Century of Spin*. In this chapter, we show to what extent international environmental policymaking in this time period was predicated on unifying styles and structures of influence through multinational corporate power, coordinated and shaped by mainly American public relations. See Levy and Newell (eds.), *The Business of Global Environmental Governance*, for representative case studies of corporate power to standardize environmental rule-making.

11. Bernstein, *Compromise*, 49.

12. Parenti, "The Limits to Growth: The Book that Launched a Movement."

13. Hecox, "Limits to Growth Revisited: Has the World Modeling Debate Made any Progress?" See also Elichirigoity, *Planet Management*.

14. Although the report had not yet been published when the conference took place, it nevertheless had a strong influence on the conference's articulation of the problem of the human environment. According to Bernstein, UNCHE organizer Maurice Strong met with MIT professors involved in the research for the report, including Jay Forrester and Donella Meadows, in January 1971. Bernstein, *Compromise of Liberal Environmentalism*, 41–42.

15. "If the present growth trends in world population, industrialization, pollution, food production, and resource depletion continue unchanged, the limits to growth on this planet will be reached sometime within the next one hundred years." *Limits to Growth*, 21.

16. Meadows et al., *Limits to Growth*, 21.

17. See Forrester, *World Dynamics*.

18. Edwards, *A Vast Machine*, 369.

19. United Nations, *An Action Plan for the Human Environment*, 13.

20. On the culture of expertise and information management around the Infoterra database, see Aronczyk, "Environment 1.0."
21. Edwards, *A Vast Machine*, 358–59.
22. Aronczyk, "Public Relations, Issue Management, and the Transformation of American Environmentalism."
23. As we saw in chapter 3, companies such as Mobil Oil explicitly designed information and influence campaigns to counter the argument in *Limits to Growth*. See also Caradonna, *Sustainability: A History.*
24. Bernstein, *Compromise.*
25. Haas, *Epistemic Communities, Constructivism, and International Environmental Politics.*
26. Cross, "The Limits of Epistemic Communities."
27. Levy and Newell, *The Business of Global Environmental Governance*, 2.
28. Levy and Newell, *The Business of Global Environmental Governance*, 2–3.
29. Stone, "Transfer Agents and Global Networks in the 'Transnationalization' of Policy," 556.
30. Stone, "Transfer Agents and Global Networks," 557. See also Aronczyk, *Branding the Nation*, for an elaboration of how consultancies promote global market expertise for national territories.
31. Hoffman, *From Heresy to Dogma.*
32. Revzin, "Brussels Babel: Europeans Are Writing the Rules Americans Will Live By." See also Revzin, "United We Stand . . ."
33. AEF, "Monitoring Project on Behalf of E. Bruce Harrison Company," 5; Doyle and May, "Europe Readies Environmental Standards."
34. In 1989, EnviroComm also helped to set up an interparliamentary network based in Brussels called the Global Legislators Organization for a Balanced Environment (GLOBE). Made up of a group of eight industrialized nations (G8), the network's mission was "to respond to urgent environmental challenges through the development and advancement of legislation." Personal interview, 2017. EnviroComm would provide input for their agendas based on their clients' issues. See "GLOBE International: History"; and "GLOBE EU 1989–1999: Ten Years of Action for the Environment."
35. Andersson Elffers Felix, Monitoring Project on Behalf of E. Bruce Harrison Company Concerning the EC Environmental Legislation Process, Utrecht/Brussels.
36. See, e.g., Harrison, "A QUALITY Approach to Environmental Communication."
37. See, e.g., an interview with the Wittenbergs on the American cable-satellite television public affairs network, C-SPAN: "Grassroots Lobbying in Washington."
38. "The Amplifier," 66. See also Kuethe, "Access as Bargaining Chip," Van Heuverswyn & Schuybroek, "Lobbying: An Old Profession Rediscovered"; Zagorin, "The Euro Peddler."
39. E.g., Wittenberg, "In Brussels, a 'Gucci Gulch.'" Throughout the 1970s and 1980s, Wittenberg ran his own public relations firm in Washington and was frequently cited in the media as an expert in image and influence, both in Washington and abroad. See, e.g., Maxa, "Image and Influence: The Public Relations Profession in Washington";

Gamarekian, "Foreign Image-Making: It's a Job for the Experts"; Gailey, "Matching Congressmen and Executives, for a Price." Wittenberg also wrote and spoke extensively about the value of public affairs for democracy. See, e.g., Wittenberg, "How Lobbying Helps Make Democracy Work," a speech delivered to the Brazilian Public Relations Congress in Brasilia on 2 September 1982.

40. Michel, "A Bruxelles Si Vous Ne Voulez Pas La Subir." ("To Brussels if You Don't Want to Suffer.")

41. United Nations Environment Programme, "World Industry Conference on Environmental Management (WICEM): Outcome and Reactions," 3. As a show of the common ground on which industry, governments, and development groups could stand, the conference featured speakers from all three sectors. William Ruckelshaus, first head of the Environmental Protection Agency in 1970, spoke of partnerships such as Clean Sites, Inc. as "bridging institutions" between industry and government as well as between industry and environmentalists. David Roderick, chair of U.S. Steel, called for a vision of environmental improvement and industrial development as "sensibly balanced—to the benefit of all and the detriment of none." The WICEM I speech by the chair of the World Commission on Environment and Development, Gro Harlem Brundtland, centered on sustainable development as a requirement for environmental protection. "There is no choice of either/or," she told the assembled delegates. "We can only achieve sustainable growth provided we manage to protect the environment and we shall only succeed in protecting the environment if we can accomplish sustainable growth."

42. US companies in attendance at WICEM I: Air Products & Chemicals, Inc., Alcoa, American Paper Institute, ARCO, Bechtel Group, The Business Roundtable, Dow Chemical Co., Du Pont, Ebara International Corp., Exxon Corp., Ford Motor Co., Gulf Oil Co., IBM, Mitchell Energy Co., Mobil Oil Corp., Monsanto, Motor Vehicle Manufacturers Association, Procter & Gamble Co., SKF Steel, Inc., Standard Oil Co., Tenneco, Inc., Texaco, Inc., 3M Center, TRW Inc., Union Oil Center, U.S. Council for International Business, U.S. Steel Corp., W. R. Grace & Co., World Environment Center. Other delegates are listed in Sallada and Doyle (eds.), *The Spirit of Versailles: The Business of Environmental Management.*

43. Founding members included Ciba-Geigy, Monsanto, Henkel, and 3M, with David Roderick of U.S. Steel as chair of its board of directors.

44. See Power, "Expertise and the Construction of Relevance," on the expansion of expertise among accountants who are increasingly called on in this time period to evaluate voluntary schemes for environmental auditing. Power sees this as part of a "managerial turn" in environmental regulation, engendering new instruments of control while destabilizing the categories of independent scientific expertise. See also Hajer, "'Verinnerlijking': The Limits to a Positive Management Approach."

45. "WICEM II: The Expected Results: First Brainstorming Session." 11–12 December 1990. Zurich. Unpublished; copy in possession of authors.

46. For a complete list of E. Bruce Harrison Co. clients, see Appendix 2.

47. "The Nairobi Code for Communication on Environment and Development," 85. See also IPRA, "View from the Gallery: A News Bulletin of the International Public Relations Association."
48. See, e.g., Popoff, "Corporate America: An Agenda for What's Right in the '90s." Frank Popoff was president and chief executive officer of the Dow Chemical Company. The PRSA formed an Environment Section in 1992, hosting conferences and publishing newsletters on topics such as "Turning Green without Getting Black and Blue," and "Environmental Stewardship: Coming of Age in the '90s," and "Smart Environmentalism," the latter billed as "public policy and private practice that's based on priorities and partnering, consensus and common sense, and, most importantly, environment-with-economics." PRSA Positioning Statement: Smart Environmentalism.
49. Harrison, "Green Communication in the Age of Sustainable Development," 5. The article, prepared in the form of a brochure for IPRA members, was funded by a grant from Alcan Smelters & Chemicals Ltd. IPRA, "View from the Gallery."
50. Harrison, "Counseling Companies on Environmental Communication."
51. EnviroComm network firms with tobacco clients include Interel (Philip Morris); Kohtes Klewes (Lucky Strikes); Sanchis (R. J. Reynolds); and Trevor Russel (Philip Morris). Based also on a report prepared by the E. Bruce Harrison Company for R. J. Reynolds advocating a European expansion of public affairs capabilities, we surmise that Harrison envisioned EnviroComm as a European platform from which to counter or weaken legislation on tobacco and/or air quality. See E. Bruce Harrison Company, "A Proposal to Serve RJR." The case of EnviroComm matches research findings on the sharing of strategic information across contentious sectors including tobacco, fossil fuels, and chemicals. See White and Bero, "Corporate Manipulation of Research"; Oreskes and Conway, Merchants of Doubt; Union of Concerned Scientists, "Smoke, Mirrors, and Hot Air"; Center for International Environmental Law, "Smoke and Fumes."
52. EnviroComm, "Franchise Agreement." Unpublished; copy in possession of authors.
53. EnviroComm Franchise Network List. Unpublished; copy in possession of authors.
54. Fortun, Advocacy after Bhopal, 63–65.
55. Garcia-Johnson, Exporting Environmentalism, 72.
56. Fortun, Advocacy after Bhopal, 65.
57. Fortun, Advocacy after Bhopal, 65.
58. EnviroComm, "Responsible Care & Environmental Community Relations."
59. EnviroComm, "Environmental Reputation Benchmarking."
60. United Nations, "Agenda 21: Programme of Action for Sustainable Development," 237. Agenda 21 was also enforced in a range of other organizations with which Harrison and/or his clients were involved: the United Nations Environment Programme (UNEP), the International Chamber of Commerce (ICC), the U.S. Council on International Business (USCIB), the Chemical Manufacturers Association (CMA), the Global Environmental Management Initiative (GEMI), the International Network for Environmental Management (INEM), and the US Business Roundtable.

61. EnviroComm, "EnviroComm Europe Issues Brief: Environmental Standards Systems Set Off a Scramble."
62. Gordon and Johnson, "The Orchestration of Global Urban Climate Governance," 708.

Chapter 6

1. Personal interview, 29 June 2017.
2. Lazarus, "Super-wicked Problems and Climate Change."
3. Manheim also promotes a dramatically ahistorical and apolitical vision, pointing to the strategic, disciplined, and dedicated nature of "information and influence campaigns" across three centuries of battles for social change—from the eighteenth-century drive to abolish the slave trade through the Nike athletic company boycotts—by which committed actors of all stripes wield "communication and action to change the behavior of another party to their advantage." Manheim, *Strategy in Information and Influence Campaigns*, 3.
4. We have changed the names of advocates and provide organization type instead of name. This was mainly at the request of our interlocutors, which is telling in and of itself. Stories of their workplace being monitored, infiltrated, or hacked were not uncommon. A full list of organizations at which interviews took place, decoupled from names, appears in Appendix 1.
5. Smith and Howe, *Climate Change as Social Drama*.
6. See, e.g., Botan and Hazleton (eds.), *Public Relations Theory II*; Grunig, ed., *Excellence in Public Relations and Communication Management*.
7. Grunig, "Review of Research on Environmental Public Relations," 46.
8. Stamm, "Conservation Communications Frontiers," 4. These concerns were equally reflected in social psychology. See, e.g., Heberlein, *Navigating Environmental Attitudes*.
9. Stamm, "Two Orientations to the Concept of Scarcity"; Stamm and Bowes, "Environmental Attitudes and Reaction"; Stamm, "Conservation Communications Frontiers."
10. Grunig, "Review of Research on Environmental Public Relations," 47.
11. As E. Bruce Harrison would explain in a presentation on grassroots campaigns to the Public Relations Society of America in May, 1987: "People are interested in themselves. They listen to messages which relate to themselves. They are moved to act when it seems important to themselves. Any coalition or interest group will hold together just so long as the individual members believe the group represents their self-interest. Messages on legislative issues should certainly identify with the 'public interest,' but they won't score well unless each individual sees in them something for himself." Harrison, "Grassroots Public Relations," 4.
12. Harrison, "Grassroots Public Relations," 49. See also Major, "Environmental Concern and Situational Communication Theory."

13. Stamm and Grunig, "Communication Situations and Cognitive Strategies in Resolving Environmental Issues"; Grunig, "Communication Behaviors and Attitudes of Environmental Publics."

14. In the urban study, those are the four questions asked. In the rural study, the last question was tweaked to be more expansive, focusing on more than "solutions": "Do you have a great deal of knowledge or experience that would help you make judgments about these issues, some knowledge or experience, very little, or none?"

15. Grunig, "A Situational Theory of Environmental Issues, Publics, and Activists," 50.

16. Grunig, "Communication Behaviors and Attitudes of Environmental Publics," 13–14.

17. Grunig, "Communication Behaviors and Attitudes of Environmental Publics," 12.

18. Dewey, The Public and Its Problems, 137.

19. Grunig's curriculum vitae gives a sense of his influence in academic and professional settings. In addition to his many years of consulting work for companies and PR firms on their PR programs and strategic planning, Grunig was a longtime advisor to the US Department of Energy on their public affairs. Among other roles, he sat on the DOE's Communication and Trust Advisory Panel, established as part of the federal agency's mea culpa after radioactive contaminants were found on the grounds and in the river surrounding their Brookhaven National Laboratory. Located 60 miles east of New York City, the Long Island laboratory's nuclear reactor was active for eighteen years (1950–1968) of DOE experiments on uses of the atom. See Cotsalas, "Brookhaven Lab's $97 Million Cleanup"; Brookhaven National Laboratory, "Institutional Plan." Over the course of his career Grunig won virtually every major award from professional public relations associations and was invited around the world to give talks on his theories and methods.

20. Grunig, "A Situational Theory of Environmental Issues, Publics, and Activists," 50–54.

21. Grunig, "A Situational Theory of Environmental Issues, Publics, and Activists," 53–54.

22. American environmental organizations are not at all alike in their orientation to business agendas and managerial techniques. Although public relations is strongly indexed to corporate power, we found in our interviews that a surprising range of environmental organizations, including those most directly opposed to business values, adopt the ideological premises of public relations, usually in the spirit of making use of the same media outlets and performative techniques to confront and redirect segmented audiences.

23. Bernstein, Compromise of Liberal Environmentalism.

24. Bosso, Environment, Inc.; Bosso and Guber, "Maintaining Presence: Environmental Advocacy and the Permanent Campaign."

25. Shellenberger and Nordhaus, "The Death of Environmentalism."

26. Mitchell, Mertig, and Dunlap, "Twenty Years of Environmental Mobilization"; Dowie, Losing Ground; Berry, The New Liberalism.

27. Brulle, Carmichael, and Jenkins argue that New York Times coverage of the release of the film An Inconvenient Truth in 2006, directed by Al Gore, boosted public perception of the urgency of climate change. See Brulle, Carmichael, and Jenkins, "Shifting Public Opinion of Climate Change." See also Boykoff, "Public Enemy No. 1?"

28. Bäckstrand and Lövbrand, "Climate Governance Beyond 2012."

29. Ganz, *Why David Sometimes Wins.*
30. See Hoggan and Ganz, "Sometimes David Wins."
31. The discussion of "truth" in climate change discourse is far more complex than what is rendered here as an opposition between truth and legitimacy. As Chris Russill reveals, claims to truth in assertions about the reality of climate change can themselves be seen as an act of strategic communication. Russill, "Truth and Opinion in Climate Change Discourse."
32. See Isaac William Martin, *Rich People's Movements*, which defines radical flank as the influence of radical protest on decisions to adopt more moderate proposals along the same lines. The idea of a radical flank comes from Herbert Haines, "Black Radicalization and the Funding of Civil Rights, 1957–1970."
33. Walker, *Grassroots for Hire*. But see Wood, "Corporate Front Groups and the Making of a Petro-Public," for an excellent account of the complex affiliations of adherents to these groups.
34. Of course, there are long-standing reasons for the antagonistic stance of some activists toward public relations. First, as we have seen throughout this book, the practice of public relations is deeply entwined with the US history of corporate power. Second, and relatedly, public relations is dominantly understood as an institutional practice and not as a set of communicative processes. As Kristin Demetrious points out, it is therefore not surprising that activists see themselves as "victims" of the "manipulative" and "undemocratic" practice of public relations and are unable or unwilling to recognize their communicative efforts in those terms. Demetrious, "Active Voices."
35. Dauvergne and LeBaron's *Protest, Inc.* (and Clifford Bob's more systematic treatment, *The Marketing of Rebellion*) presumes a "corporatization" of activism, through branding, institutionalization and fundraising. Dauvergne and LeBaron describe this process as a capitulation by activists to a privatized, consumption-based set of practices rooted in neoliberal market norms. Our research suggests that the terrain is far more complex. Our interviewees were well aware of the spirit of compromise characterizing the climate of publicity into which they entered and the implications of the choices they make.
36. To the extent that self-interest is often a prime motivation for participation in deliberative processes, "this has opened the door for all kinds of organizations to be recognized as deliberative agents . . . and for PR, as the organizational function through which deliberation is institutionally managed, to be recognized as an important influence on deliberative engagements." Edwards, *Understanding Public Relations*, 88.
37. See, e.g., Callison, *How Climate Change Comes to Matter*; Lippert, "Corporate Carbon Footprinting as Techno-Political Practice"; Lohmann, "Marketing and Making Carbon Dumps"; Pasek, "Managing Carbon and Data Flows."
38. Cronon, "The Trouble with Wilderness," 21.

Chapter 7

1. Porter and Kramer, "Creating Shared Value," 71–75; see also Porter and Kramer, "Strategy and Society: The Link Between Competitive Advantage and Corporate Social Responsibility."
2. On the information deficit argument around climate change, see Bulkeley, "Common Knowledge?" On uses of mobile data for social good, see, e.g., Poom et al., "COVID-19 is Spatial."
3. See The Future Society, "AI4SDG: Roadmap to a Global Data Commons to Achieve the Sustainable Development Goals."
4. Yakowitz, "Tragedy of the Data Commons," 4.
5. Lippert, "Failing the Market, Failing Deliberative Democracy"; Lippert, "Corporate Carbon Footprinting as Techno-Political Practice"; Vesty, Telegenkamp, and Roscoe, "Creating Numbers."
6. Fortun, "Environmental Information Systems as Appropriate Technology"; Gabrys, *Program Earth. Environmental Sensing Technology and the Making of a Computational Planet*; Mah, "Environmental Justice in the Age of Big Data."
7. Levy, "Environmental Management as Political Sustainability"; Power, "Expertise and the Construction of Relevance"; Levy and Newell, *The Business of Global Environmental Governance*.
8. Levy, "Environmental Management," 127.
9. Collaborators in D4CA and related initiatives include other United Nations bodies such as the UN Development Operations Coordination Office; economic organizations such as the World Economic Forum; national governments; sustainability research groups such as the World Business Council for Sustainable Development; and private data companies such as mobile network operators.
10. Fortun, "Environmental Information Systems as Appropriate Technology," 54.
11. Fortun, "From Bhopal to the Informating of Environmentalism."
12. Kuneva, "Keynote Speech: Roundtable on Online Data Collection, Targeting and Profiling." Emphasis added.
13. World Economic Forum, "Personal Data: The Emergence of a New Asset Class."
14. Hajer et al., "Beyond Cockpit-ism"; World Economic Forum, "Rethinking Personal Data: A New Lens for Strengthening Trust."
15. Couldry & Meijias, "Data Colonialism," 340.
16. Fourcade and Kluttz call this a "Maussian bargain," by which "the harvesting of data about people, organizations and things and their transformation into a form of capital" is made to seem not as dispossession but as a benign process of gift-like exchange, by which all parties to the exchange stand to benefit in material and symbolic ways. Fourcade and Kluttz, "A Maussian Bargain."
17. Couldry and Meijias, "Data Colonialism," 340.
18. Isin and Ruppert, "Data's Empire: Postcolonial Data Politics." Isin and Ruppert also draw on United Nations Global Pulse initiatives to make their case.
19. Isin and Ruppert, "Data's Empire." See also Scott, *Seeing Like a State*; Anderson, *Imagined Communities*.

20. Power examines the emergence of environmental accounting techniques in the UK in the mid-1990s, by which the jurisdiction of environmental concerns is enlarged to accommodate professional accounting language, practices, and expertise. By focusing on the claims to legitimacy as opposed to investigating the actual legitimacy of environmental auditing, Power allows us to apprehend the performative dimensions of the various representational strategies in which environmental auditors engage as well as the making of private-sector expertise in dealing with matters of environmental concern. Power, "Expertise and the Construction of Relevance."

21. Lippert's work on carbon accounting recognizes the co-constitution of such markets by a range of actors. Not only corporate firms but also organizations such as the Organization for Economic Cooperation and Development (OECD) and the World Business Council for Sustainable Development (WBCSD) are involved in the legitimation of global carbon reports and metrics. Lippert, "Corporate Carbon Footprinting as Techno-Political Practice." See also Lohmann, "Marketing and Making Carbon Dumps"; Williams, Whiteman, and Parker, "Backstage Interorganizational Collaboration."

22. Pasek, "Managing Carbon and Data Flows." See also Russill, "Looking for the Horizon," who argues that media theory and environmental science can be constituted through the same problematics.

23. Lazarus, "Super-wicked Problems and Climate Change."

24. Faghmous and Kumar, "A Big Data Guide to Understanding Climate Change," 16.

25. Fortun, "Environmental Information Systems as Appropriate Technology"; Lippert, "Failing the Market, Failing Deliberative Democracy," 2; Gabrys, *Program Earth*; Aronczyk, "Environment 1.0: Infoterra and the Making of Environmental Information."

26. Gabrys, "Practicing, Materializing and Contesting Environmental Data"; Gabrys, "The Becoming Environmental of Computation"; Mah, "Environmental Justice in the Age of Big Data."

27. But even in the most fair-minded, participatory, and publicly accessible contexts, many of them still rely on corporate information and technology infrastructures (e.g., the Google Maps platform) for data collection and analysis. See Mah, "Environmental Justice in the Age of Big Data."

28. Bloomberg Finance LP, "Data for Good Exchange 2018." Emphasis in original.

29. Aakus and Bzdak, "Revisiting the Role of 'Shared Value' in the Business-Society Relationship"; Porter and Kramer, "Strategy and Society"; "Creating Shared Value."

30. World Economic Forum, "Rethinking Personal Data" (2010); World Economic Forum, "Personal Data: The Emergence of a New Asset Class" (2011); UN Global Pulse, "Big Data for Development: Challenges and Opportunities" (2012); World Economic Forum, "Rethinking Personal Data: A New Lens for Strengthening Trust" (2014); UN Secretary-General Independent Expert Advisory Group, "A World that Counts: Mobilizing the Data Revolution for Sustainable Development" (2014); World Economic Forum, "Paving the Path to a Big Data Commons" (2015).

31. World Economic Forum, "Personal Data: The Emergence of a New Asset Class."

32. UN Secretary-General Independent Expert Advisory Group, "A World that Counts."

33. World Economic Forum, "Personal Data: The Emergence of a New Asset Class."
34. World Economic Forum, "Rethinking Personal Data: A New Lens for Strengthening Trust," 3; 33.
35. Fortun, "Informating Environmentalism."
36. Wood and Aronczyk, "Publicity and Transparency."
37. World Economic Forum, "Personal Data: The Emergence of a New Asset Class."
38. Giovanni and Jespersen, "You Say You Want a Revolution"; Jahan, "The Data Revolution for Human Development."
39. UN Secretary-General Independent Expert Advisory Group, "A World that Counts," 27.
40. UN Global Pulse, "Big Data for Development."
41. Kirkpatrick, "Data Philanthropy: Public & Private Sector Data Sharing for Global Resilience."
42. Hajer et al., "Beyond Cockpit-ism," 1652. It is relevant that some of the authors of this article are based at the World Business Council for Sustainable Development and the Stockholm Resilience Center, organizations which since the early 2000s have aimed to interpret climate change through the lens of risk management and, as Lippert ("Corporate Carbon Footprinting") indicates, are part of the co-constitution of voluntary, private regimes of environmental response such as corporate carbon accounting.
43. Hajer et al., "Beyond Cockpit-ism," 1656.
44. Hajer et al., "Beyond Cockpit-ism," 1658.
45. Kirkpatrick, "Unpacking the Issue of Missed Use and Misuse of Data."
46. Biruk, *Cooking Data.*
47. Crystal Biruk, among others, has noted how this is often the "default language" for conceptualizing the link between action and development research worlds. Biruk, *Cooking Data*, 168.
48. See, e.g., Biruk, *Cooking Data*; Green, "Calculating Compassion"; Mosse, *Cultivating Development.*
49. UN ECOSOC, "Partnering for Resilient and Inclusive Societies."
50. *UN News,* "UN Makes 'Declaration of Digital Interdependence.'"
51. UN Secretary-General, "The Age of Digital Interdependence."
52. Rajpurohit, "Interview: Emanuel Letouzé, Data-Pop Alliance on Big Data and Human Rights."
53. Letouzé and Vinck, "A New and Sometimes Awkward Relationship."

Conclusion

1. For a list of transformations to US environmental and science policy under the Trump administration during its first two years in office, see *National Geographic,* "A Running List." For 2019–2020 changes, see McKeever, "Trump's and Biden's Policy Promises and Actions." For accounts of efforts by the Trump administration to suppress or

destroy information and federal datasets pertaining to environmental protection, see Dillon et al., "Environmental Data Justice and the Trump Administration"; and Russell and Tegelberg, "Beyond the Boundaries of Science."

2. Peterson Companies, "Waterfront Development."
3. Macdonald, "National Harbor a Threat to the Potomac"; and Byrne, "National Harbor: And the Environment?" The quoted passages are from Byrne.
4. Williams, "Return from the Nadir." For a description of the ambit of Public Interest Research Groups today, see USPIRG.org.
5. See, e.g., Levick, "The Interview (Richard S. Levick)."
6. Levick, "Insights in New Media and Public Relations," keynote speech, Oil & Gas Public Relations and New Media Conference, National Harbor, Maryland, 7 May 2015. The TransCanada Corporation's Keystone XL pipeline was a controversial pipeline project intended to carry over 800,000 barrels of crude oil per day from Canada to the US Gulf Coast. In 2015, when Levick gave this speech, the pipeline project was mired in a lengthy review process (and would be rejected by President Obama later that year, citing concerns over carbon emissions). Regulations were removed by the Trump administration, allowing construction of the pipeline to continue; but in January, 2021, newly elected US president Joe Biden revoked the Keystone XL pipeline permit on his first day in office.
7. Holt, "Shifting Paradigms: Broadening the Discussion on O&G Development." Speech delivered at Oil & Gas Public Relations and New Media Conference, National Harbor, Maryland, 8 May 2015. David Holt is also the founder and managing partner of HBW Resources, a public relations firm and strategic consultancy with offices throughout the United States. HBW Resources created the Consumer Energy Alliance (CEA) in 2006 to promote oil, gas, and tar sands infrastructure and development. The CEA, which maintains operations in twenty states, is housed in HBW Resources offices.
8. One public event described was the Consumer Energy Alliance sponsorship of an annual Energy Day Festival in Houston, Texas, a "family event" designed to educate ordinary citizens about the benefits of the energy industry. Such public events harken back to the public and community relations of the Chemical Manufacturing Association starting in the 1950s (see chapter 3).
9. In his own speech, Richard Levick made this claim as well, asking attendees, "How many people here are anti-environment?" and letting the silence answer his question.
10. We are grateful to Tim Wood for helping us to formulate this point.
11. Lee et al., Democratizing Inequalities.
12. Pasek, "Mediating Climate, Mediating Scale."
13. See, e.g., Yosie and Herbst, "Using Stakeholder Processes in Environmental Decisionmaking"; and Yosie, "Emerging Strategies to Manage System-Level Risks." A former director of the US Environmental Protection Agency's Science Advisory Board, then vice-president of health and environment at the American Petroleum Institute, Terry Yosie joined the staff of the E. Bruce Harrison Company in 1992, becoming the PR firm's top analyst in the strategic management of federal environmental and health policy to promote industrial benefits and mitigate the policy impacts of scientific evidence.

14. Russill, "Dewey/Lippmann Redux," 130.
15. We are influenced in some measure by the work of Bruno Latour (e.g., *An Inquiry into Modes of Existence* and "From *Realpolitik* to *Dingpolitik*"), Noortje Marres ("Issues Spark a Public into Being," and "The Issues Deserve More Credit") and their collaborators on the problem-solving potential of John Dewey and Walter Lippmann's conceptions of democracy to rethink the relationship of science and technology to society. In particular, Latour's proposal of a turn from "matters of fact" to "matters of concern" parallels to some extent the argument we are making here. Yet we want also to conserve the historical arguments made by Dewey and Lippmann, and later Hannah Arendt, in their reckoning with concepts of truth and politics in the development of a historical consciousness. See also Russill, "Dewey/Lippmann Redux," on this point; and see the essays in Arendt, *Between Past and Future*.

Bibliography

Archives and Libraries

H. John Heinz III Collection, 1971–1991, Carnegie Mellon University Digital Collection. https://digitalcollections.library.cmu.edu

Papers of the Manufacturing Chemists Association (later Chemical Manufacturers Association), Chemical Industry Archives, University of California-San Francisco Library. https://www.industrydocuments.ucsf.edu/chemical/

Maurice Strong Papers, Environmental Science and Public Policy Archives (ESSPA), Harvard University, Harvard College Library, Cambridge, MA

E. Bruce Harrison Company Papers, personal collection, in possession of authors

George H.W. Bush Presidential Library. https://bush41library.tamu.edu

Greenpeace Investigations. https://research.greenpeaceusa.org

Museum of Public Relations Collection, Baruch College, New York, NY

Office of International Programs Records, 1964–1976; Smithsonian Science Information Exchange Records, 1946–1983, Smithsonian Institution Archives, Washington, DC

Truth Tobacco Industry Documents, University of California-San Francisco Library. https://www.industrydocuments.ucsf.edu/tobacco/

Public Relations Society of America Records (PRSA), 1983–2013, Wisconsin Historical Society, Madison, WI

Published works

"A Plan to Save the Forests: Forest Preservation by Military Control." *Century Magazine* 49 (1895): 626–634. https://babel.hathitrust.org/cgi/pt?id=ucl.31822019654656&view=1up&seq=638.

Aakhus, Mark, and Michael Bzdak. "Revisiting the Role of 'Shared Value' in the Business-Society Relationship." *Business & Professional Ethics Journal* 31, no. 2 (2012): 231–246.

Adams, Henry C. "What Is Publicity?" *North American Review* 175, no. 553 (December 1902): 895–904.

American Forestry Association. "Proceedings of the American Forest Congress." Washington, DC: H. M. Suter, 1905. https://babel.hathitrust.org/cgi/pt?id=ucl.$b11433&view=1up&seq=5.

Anderson, Benedict. *Imagined Communities*. London: Verso, 1991.

Anderson, Chris. "The End of Theory: The Data Deluge Makes the Scientific Method Obsolete." *Wired Magazine*, 23 June 2008.

Anderson, Jack, and Les Whitten. "The Washington Merry-Go-Round: Chile Resorts to Book Burning." *Washington Post*, 30 August 1975.

Andersson Elffers Felix. "Monitoring Project on Behalf of E. Bruce Harrison Company Concerning the EC Environmental Legislation Process." Utrecht/Brussels: AEF, 1989.

Andrews, Thomas G. "The Road to Ludlow: Work, Environment, and Industrialization, 1870–1915." PhD dissertation, University of Wisconsin-Madison, 2003.

Apthorpe, Raymond, and Des Gasper, eds. *Arguing Development Policy: Frames and Discourses*. London: Routledge, 1996.

Arendt, Hannah. "Truth and Politics." *New Yorker*, 25 February 1967.

Arendt, Hannah. *Between Past and Future*. New York: Penguin Random House, [1961] 2006.

Aronczyk, Melissa. "Environment 1.0: Infoterra and the Making of Environmental Information." *New Media and Society* 20, no. 5 (2017): 1832–1849.

Aronczyk, Melissa. "'Living the Brand': Nationality, Globality and the Identity Strategies of Nation Branding Consultants." *International Journal of Communication* 2 (2008): 41–65.

Aronczyk, Melissa. "Understanding the Impact of the Transnational Promotional Class on Political Communication." *International Journal of Communication* 9, no. 1 (2015): 2007–26.

Aronczyk, Melissa. "Public Relations, Issue Management, and the Transformation of American Environmentalism, 1948–1992." *Enterprise & Society* 18, no. 4 (2018): 836–863.

Aronczyk, Melissa. "Environment 1.0: Infoterra and the Making of Environmental Information." *New Media and Society* 20, no. 5 (2018): 1832–1849.

Auerbach, Jonathan. *Weapons of Democracy: Propaganda, Progressivism, and American Public Opinion*. Baltimore: Johns Hopkins University Press, 2015.

Awad, Joseph F. "Environment: A Continuing Arena." *Public Relations Journal* (May 1973): 2.

Bäckstrand, Karen, and Eva Lövbrand. "Climate Governance Beyond 2012: Competing Discourses of Green Governmentality, Ecological Modernization, and Civic Environmentalism." In *The Social Construction of Climate Change*, ed. Mary E. Pettenger, 123–148. London: Routledge, 2007.

Barenberg, Mark. "Democracy and Domination in the Law of Workplace Cooperation: From Bureaucratic to Flexible Production." *Columbia Law Review* 94, no. 3 (1994): 753–983.

Barley, Stephen R. "Building an Institutional Field to Corral a Government: A Case to Set an Agenda for Organization Studies." *Organization Studies* 31, no. 6 (2010): 777–805.

Bartley, Tim. "How Foundations Shape Social Movements: The Construction of an Organizational Field and the Rise of Forest Certification." *Social Problems* 54, no. 3 (2007): 229–255.

Beder, Sharon. *Global Spin: The Corporate Assault on Environmentalism*. Cambridge: Green Books, 1998.

"Begin Fight to Save the Yosemite Park." *New York Times*, 12 January 1909. https://www.nytimes.com/1909/01/12/archives/begin-fight-to-save-the-yosemite-park-historic-and-scenic.html.

Bennett, W. Lance, and Shanto Iyengar. "A New Era of Minimal Effects? The Changing Foundations of Political Communication." *Journal of Communication* 58, no. 1 (2008): 707–731.

Berger, Peter L., and Thomas Luckmann. *The Social Construction of Reality: A Treatise in the Sociology of Knowledge*. New York: Penguin Books, 1966.

Bernays, Edward L. "Manipulating Public Opinion: The Why and the How." *American Journal of Sociology* 33, no. 6 (1928): 958–971.

Bernstein, Steven. *The Compromise of Liberal Environmentalism.* New York: Columbia University Press, 2001.

Berry, Jeffrey M. *Lobbying for the People: The Political Behavior of Public Interest Groups.* Princeton, NJ: Princeton University Press, 1977.

Berry, Jeffrey M. *The New Liberalism: The Rising Power of Citizen Groups.* Washington, DC: Brookings Institution Press, 1999.

Bessy, Christian, and Pierre-Marie Chauvin. "The Power of Market Intermediaries: From Information to Valuation Processes." *Valuation Studies* 1, no. 1 (2013): 83–117.

Best, George E. "A Rational Approach to Air Pollution Legislation." *American Industrial Hygiene Association Quarterly* 13, no. 2 (1952): 62–69.

Biruk, Crystal. *Cooking Data: Culture and Politics in an African Research World.* Durham, NC: Duke University Press, 2018.

Björk, Tord. "The Emergence of Popular Participation in World Politics—United Nations Conference on Human Environment 1972," University of Stockholm, 1996.

Blumenthal, Frank H. "Anti-Union Publicity in the Johnstown 'Little Steel' Strike of 1937." *Public Opinion Quarterly* 3, no. 4 (1939): 676–682.

Bob, Clifford. *The Marketing of Rebellion: Insurgents, Media, and International Activism.* Cambridge: Cambridge University Press, 2010.

Bocking, Stephen. *Nature's Experts: Science, Politics, and the Environment.* New Brunswick, NJ: Rutgers University Press, 2004.

Bogner, Alexander, Beate Littig, and Wolfgang Menz, eds. *Interviewing Experts.* London: Palgrave Macmillan, 2009.

Bollig, Michael. "Resilience—Analytical Tool, Bridging Concept or Development Goal? Anthropological Perspectives on the Use of a Border Object." *Zeitschrift Für Ethnologie* 139, no. 2 (2014): 253–279.

Boltanski, Luc, and Laurent Thévenot. *On Justification: Economies of Worth.* Princeton, NJ: Princeton University Press, 2006.

Bosso, Christopher J. *Environment, Inc.: From Grassroots to Beltway.* Lawrence: University Press of Kansas, 2005.

Bosso, Christopher J., and Deborah Lynn Guber. "Maintaining Presence: Environmental Advocacy and the Permanent Campaign," In *Environmental Policy: New Directions for the Twenty-first Century*, ed. Norman J. Vig and Michael E. Kraft. 78–99. Washington, DC: CQ Press, 2006.

Botan, Carl H., and Vincent Hazleton, eds. *Public Relations Theory II.* London: Routledge, 2006.

Bourdieu, Pierre. "Opinion Polls: A 'Science' Without a Scientist." In *In Other Words: Essays Toward a Reflexive Sociology*, 168–176. Stanford, CA: Stanford University Press, 1990.

Bowker, Geoffrey C. *Science on the Run: Information Management and Industrial Geophysics at Schlumberger, 1920-1940.* Cambridge, MA: MIT Press, 1994.

Bowker, Geoffrey C., Karen Baker, Florence Miller, and David Ribes. "Toward Information Infrastructure Studies: Ways of Knowing in a Networked Environment." *International Handbook of Internet Research* (2009), 97–117.

Boykoff, Max. "Public Enemy No. 1? Understanding Media Representations of Outlier Views on Climate Change." *American Behavioral Scientist* 57, no. 6 (2013): 796–817.

Bradley, Joseph W. *Role of Trade Associations and Professional Business Societies in America.* University Park: Pennsylvania State University Press, 1965.

Brandt, Ellis N. "Wanted: Environmentalists." *Public Relations Journal* (August 1970): 19–21.

Brodie, Patrick. "Climate Extraction and Supply Chains of Data." *Media, Culture and Society* 42, no. 7–8 (2020): 1095–1114.

Brookhaven Science Associates. "Institutional Plan, FY 2001–FY 2005." Brookhaven National Laboratory, October 2000.

Broome, André, and Joel Quirk. "Governing the World at a Distance: The Practice of Global Benchmarking." *Review of International Studies* 41, no. 5 (2015): 819–841.

Brown, Clyde, and Herbert Waltzer. "Buying National Ink: Advertorials by Organized Interests in TIME Magazine, 1985–2000." *Journal of Political Marketing* 5, no. 4 (2007): 19–45.

Brown, Clyde, and Herbert Waltzer. "Every Thursday: Advertorials by Mobil Oil on the Op-Ed Page of the *New York Times*." *Public Relations Review* 31, no. 1 (2005): 197–208.

Brown, Halina Szejnwald, W. Martin de Jong, and Teodorina Lessidrenska. "The Rise of the Global Reporting Initiative: A Case of Institutional Entrepreneurship." *Environmental Politics* 18, no. 2 (2009): 182–200.

Brulle, Robert J. *Agency, Democracy, and Nature: The U.S. Environmental Movement from a Critical Theory Perspective.* Cambridge, MA: MIT Press, 2000.

Brulle, Robert J. "Institutionalizing Delay: Foundation Funding and the Creation of U.S. Climate Change Counter-Movement Organizations." *Climatic Change* 122, no. 4 (2014): 681–694.

Brulle, Robert J., and Robert D. Benford. "From Game Protection to Wildlife Management: Frame Shifts, Organizational Development, and Field Practices." *Rural Sociology* 77, no. 1 (2012): 62–88.

Brulle, Robert J., Jason Carmichael, and J. Craig Jenkins. "Shifting Public Opinion on Climate Change." *Climatic Change* 114 (2012): 169–188.

Buchholz, Rogene A., Alfred A. Marcus, and James E. Post. *Managing Environmental Issues: A Casebook.* Englewood Cliffs, NJ: Prentice Hall, 1992.

Buell, Lawrence. "Toxic Discourse." *Critical Inquiry* 24, no. 3 (1998): 639–665.

Bulkeley, Harriet. "Common Knowledge? Public Understanding of Climate Change in Newcastle, Australia." *Public Understanding of Science* 9 (2000): 313–333.

Bulkeley, Harriet, Liliana B. Andonova, Michele M. Betsill, Daniel Compagnon, Thomas Hale, Matthew J. Hoffmann, Peter Newell, Matthew Paterson, Charles Roger, and Stacy D. VanDeveer. *Transnational Climate Change Governance.* New York: Cambridge University Press, 2014.

Byrne, Jeb. "National Harbor: And the Environment?" Letter to the Editor. *Washington Post*, 23 January 2000.

Cadwalladr, Carole. "Cambridge Analytica a Year on: 'A Lesson in Institutional Failure.'" *The Guardian*, 17 March 2019.

Calfee, Christopher H. "Europe's 'Jolly Green Giant': Environmental Policy in the European Union." *Environs* 22, no. 1 (1998): 45–58.

Calhoun, Craig, ed. *Habermas and the Public Sphere.* Cambridge: MIT Press, 1993.

Callison, Candice. *How Climate Change Comes to Matter.* Durham, NC: Duke University Press, 2014.

Callon, Michel, ed. *The Laws of the Markets.* Oxford: Blackwell, 1998.

Caradonna, Jeremy L. *Sustainability: A History.* Oxford: Oxford University Press, 2014.

Carmichael, Jason T., J. Craig Jenkins, and Robert J. Brulle. "Building Environmentalism: The Founding of Environmental Movement Organizations in the United States, 1900–2000." *Sociological Quarterly* 53, no. 3 (2012): 422–453.

Carroll, James D. "Participatory Technology." *Science* 171, no. 1 (1971): 647–653.

Center for International Environmental Law. "Smoke and Fumes: A Hidden History of Oil and Tobacco," 2016. https://www.smokeandfumes.org/.

Chase, W. Howard. *Issue Management: Origins of the Future*. Stamford, CT: Issue Action Publications, 1985.

"Cleaning Up the Clean Air Act: National Clean Air Coalition." *Environment: Science and Policy for Sustainable Development* 23, no. 6 (1981): 16, 20, 42–44.

Clemens, Elisabeth S. *The People's Lobby: Organizational Innovation and the Rise of Interest Group Politics in the United States, 1890–1925*. Chicago: University of Chicago Press, 1997.

Cohen, Richard. "The Business Lobby Discovers That in Unity There Is Strength." *National Journal* 28 (1980): 1050–1055.

Coll, Steve. *Private Empire: ExxonMobil and American Power*. New York: Penguin Books, 2013.

Conley, Joe Greene. "Environmentalism Contained: A History of Corporate Responses to the New Environmentalism." PhD dissertation, Princeton University, 2006.

"Coordination with the United Nations System: WHO's Human Health and Environment Programme." Twenty-eighth World Health Assembly, Geneva, Switzerland. Provisional Agenda Item 3.16.6. A28/27, 15 April 1975.

Cotsalas, Valerie. "Brookhaven Lab's $97 Million Cleanup." *New York Times*, 24 April 2005.

Couldry, Nick, and Ulises A. Mejias. "Data Colonialism: Rethinking Big Data's Relation to the Contemporary Subject." *Television and New Media* 20, no. 4 (2019): 336–349.

Cronon, William. *Nature's Metropolis: Chicago and the Great West*. New York: W. W. Norton, 1992.

Cronon, William. "The Trouble with Wilderness; or, Getting Back to the Wrong Nature." In *Uncommon Ground: Rethinking the Human Place in Nature*, ed. William Cronon, 69–90. New York: W. W. Norton, 1995.

Cross, Mai'a K. Davis. "Rethinking Epistemic Communities Twenty Years Later." *Review of International Studies* 39, no. 1 (2013): 137–160.

Cross, Mai'a K. Davis. "The Limits of Epistemic Communities: EU Security Agencies." *Politics and Governance* 3, no. 1 (2015): 90–100.

C-SPAN. "Grassroots Lobbying in Washington," 12 December 1989. https://www.c-span.org/video/?10259-1/grassroots-lobbying-washington.

Cutlip, Scott M. *The Unseen Power: Public Relations, A History*. Hillsdale, NJ: Lawrence Erlbaum Associates, 1994.

"D.C. Agency Created First Client." *Publicist*, March/April (1982): 1–4.

Dadush, Sarah. "Regulating Social Finance: Can Social Stock Exchanges Meet the Challenge?" *University of Pennsylvania Journal of International Law* 37, no. 1 (2015): 139–228.

Dauvergne, Peter, and Geneviève LeBaron. *Protest, Inc.: The Corporatization of Activism*. Cambridge, UK: Polity, 2014.

Davidson, Debra J., and Mike Gismondi. *Challenging Legitimacy at the Precipice of Energy Calamity*. New York: Springer, 2011.

Davies, William. "How Statistics Lost Their Power—and Why We Should Fear What Comes Next." *The Guardian*, 19 January 2017.

Daymon, Christine, and Kristin Demetrious, eds. *Gender and Public Relations: Critical Perspectives on Voice, Image and Identity*. New York: Routledge, 2016.

DeLuca, Kevin Michael. *Image Politics: The New Rhetoric of Environmental Activism*. New York: Routledge, 2005.

Demetrious, Kristin. *Public Relations, Activism, and Social Change: Speaking Up*. New York: Routledge, 2013.

Demetrious, Kristin. "Active Voices." In *Public Relations: Critical Debates and Contemporary Practice*, ed. J. L'Etang, and M. Pieczka, 93–107. London: Lawrence Erlbaum, 2006.

Dennehy, Kevin. "First Forester: The Enduring Conservation Legacy of Gifford Pinchot." New Haven: Yale School of the Environment, 2016. https://environment.yale.edu/news/article/first-forester-the-conservation-legacy-of-gifford-pinchot/.

Dewey, John. *The Public and Its Problems*. Athens: Swallow Press/Ohio University Press, [1927] 1991.

Dillon, Lindsey, Dawn Walker, Nicholas Shapiro, Vivian Underhill, Megan Martenyi, Sara Wylie, Rebecca Lave, Michelle Murphy, Phil Brown, and Environmental Data and Governance Initiative. "Environmental Data Justice and the Trump Administration: Reflections from the Environmental Data and Governance Initiative." *Environmental Justice* 10, no. 6 (2017): 186–192.

Domhoff, G. William. "The Rise and Fall of Labor Unions in the U.S. from the 1830s until 2012 (but Mostly the 1930s–1980s)." 2013. https://whorulesamerica.ucsc.edu/power/history_of_labor_unions.html.

Dowie, Mark. *Losing Ground: American Environmentalism at the Close of the Twentieth Century*. Cambridge, MA: MIT Press, 1995.

Downie, Christian. "Fighting for King Coal's Crown: Business Actors in the US Coal and Utility Industries." *Global Environmental Politics* 17, no. 1 (2017): 21–39.

Doyle, Julian, and Susan May. "Europe Readies Environmental Standards." *Financial Executive* 7, no. 5 (September–October 1991): 53–61.

Drew, Elizabeth. "A Reporter at Large: Conversation with a Citizen." *New Yorker*, July 1973.

Dunaway, Finis. *Seeing Green: The Use and Abuse of American Environmental Images*. Chicago: University of Chicago Press, 2018.

Dunlap, Riley E., and Aaron M. McCright. "Organized Climate Change Denial." In *The Oxford Handbook of Climate Change and Society*, ed. John S. Dryzek, Richard B. Norgaard, and David Schlosberg. Oxford: Oxford University Press, 2011.

Dunlap, Thomas. *DDT: Scientists, Citizens, and Public Policy*. Princeton, NJ: Princeton University Press, 2014.

Dunlop, Claire A. "The Irony of Epistemic Learning: Epistemic Communities, Policy Learning and the Case of Europe's Hormones Saga." *Policy and Society* 36, no. 2 (2017): 215–232.

E. Bruce Harrison Company. "A Proposal to Serve RJR: ETS Strategies." Washington, DC: E. Bruce Harrison Company, 1994. https://www.industrydocuments.ucsf.edu/docs/#id=lfkv0087.

E. Bruce Harrison Company. "Grassroots Involvement: Key to Issue Management." PRSA Records, Box 156, 1983.

Edelman, Murray J. *The Politics of Misinformation*. Cambridge: Cambridge University Press, 2001.

Edwards, Lee. "Defining the 'Object' of Public Relations Research: A New Starting Point." *Public Relations Inquiry* 1, no. 1 (2012): 7–30.

Edwards, Lee. *Power, Diversity, and Public Relations*. New York: Routledge, 2015.

Edwards, Lee. "The Role of Public Relations in Deliberative Systems." *Journal of Communication* 66, no. 1 (2016): 60–81.

Edwards, Lee. *Understanding Public Relations: Theory, Culture & Society*. London: Sage, 2018.

Edwards, Lee, and Caroline E. M. Hodges. "Implications of a (Radical) Socio-Cultural 'Turn' in Public Relations Scholarship." In *Public Relations, Society and Culture: Theoretical and Empirical Explorations*, ed. Lee Edwards and Caroline E. M. Hodges, 1–14. London: Routledge, 2011.

Edwards, Lee, and Caroline E. M. Hodges, eds. *Public Relations, Society & Culture: Theoretical and Empirical Explorations*. London: Routledge, 2011.

Edwards, Paul N. *A Vast Machine: Computer Models, Climate Data, and the Politics of Global Warming*. Cambridge, MA: MIT Press, 2010.

Elichirigoity, Fernando. *Planet Management: Limits to Growth, Computer Simulation, and the Emergence of Global Spaces*. Chicago: Northwestern University Press, 1999.

EnviroComm International. *Environmental Reputation Benchmarking: A Business Development Aid for the Exclusive Use of EnviroComm Practitioners*. Washington, DC: EnviroComm International, 1995.

EnviroComm International. *Environmental Standards Systems Set Off a Scramble*. Washington, DC: EnviroComm International, 1995.

EnviroComm International. *Responsible Care & Environmental Community Relations: A Business Development Aid for the Exclusive Use of EnviroComm Practitioners*. Washington, DC: EnviroComm International, 1995.

"Environmental Partnerships Help Business Find Effective Solutions." *Business and the Environment* 3, no. 13 (1992).

Erskine, Hazel. "The Polls: Pollution and Industry." *Public Opinion Quarterly* 36, no. 2 (1972): 263–280.

Erskine, Hazel. "The Polls: Pollution and Its Costs." *Public Opinion Quarterly* 36, no. 1 (1972): 120–135.

Eulau, Heinz. "Man Against Himself: Walter Lippmann's Years of Doubt." *American Quarterly* 4, no. 4 (1952): 291–204.

Ewen, Stuart. *PR! A Social History of Spin*. New York: Basic Books, 1996.

Faghmous, James H., and Vipin Kumar. "A Big Data Guide to Understanding Climate Change: The Case for Theory-Guided Data Science." *Big Data* 2, no. 3 (2014): 155–163.

Fears, Darryl, and Steven Mufson, "Liberal, Progressive—and Racist?" *Washington Post*, 22 July 2020.

Ferguson, James. *Give a Man a Fish: Reflections on the New Politics of Distribution* Durham, NC: Duke University Press, 2015.

Ferrari, Michelle (dir.). *Rachel Carson* [film]. Boston, MA: WGBH Educational Foundation, 2017.

Fitch, Kate. "'The PR Girl': Gender and Embodiment in Public Relations." In *Popular Culture and Social Change: The Hidden Work of Public Relations*, ed. Kate Fitch and Judy Motion. London: Routledge, 2020.

Ford, Rochelle, and Cedric Brown. "State of the PR Industry: Defining and Delivering on the Promise of Diversity." National Black Public Relations Society, 2015. https://www.odwyerpr.com/site_images/NBPRS-State-of-the-PR-Industry-White-Paper.pdf.

Forrester, Jay W. *World Dynamics*. Cambridge: Wright-Allen Press, 1971.

Fortun, Kim. *Advocacy after Bhopal: Environmentalism, Disaster, New Global Orders*. Chicago: University of Chicago Press, 2001.

Fortun, Kim. "Environmental Information Systems as Appropriate Technology." *Design Issues* 20, no. 3 (2004): 54–65.

Fortun, Kim. "From Bhopal to the Informating of Environmentalism: Risk Communication in Historical Perspective." *Osiris* 19, no. 2 (2004): 283–296.

Fortun, Kim. "Biopolitics and the Informating of Environmentalism." In *Lively Capital: Biotechnologies, Ethics, and Governance in Global Markets*, ed. Kaushik Sunder Rajan, 306–326. Durham: Duke University Press, 2012.

Fourcade, Marion, and Kieran Healy. "Seeing like a Market." *Socio-Economic Review* 15, no. 1 (2017): 9–29.

Fourcade, Marion, and Daniel N. Kluttz. "A Maussian Bargain: Accumulation by Gift in the Digital Economy." *Big Data & Society* 7, no. 1 (2020): 1–16.

Fox, Stephen R. *John Muir and His Legacy: The American Conservation Movement*. Boston: Little, Brown, 1981.

Freed, Bruce. "Melding PR and Lobbying Impact." *Impact* (February 1992): 1, 5.

Freudenburg, William R., and Margarita Alario. "Weapons of Mass Distraction: Magicianship, Misdirection, and the Dark Side of Legitimation." *Sociological Forum* 22, no. 2 (2007): 146–173.

Future Society. "AI4SDG: Roadmap to a Global Data Commons to Achieve the Sustainable Development Goals." https://thefuturesociety.org/wp-content/uploads/2019/08/GDC-poster.pdf.

Gabrys, Jennifer. "Practicing, Materialising and Contesting Environmental Data." *Big Data & Society* 3, no. 2 (2016): 1–7.

Gabrys, Jennifer. *Program Earth: Environmental Sensing Technology and the Making of a Computational Planet*. Minneapolis: University of Minnesota Press, 2016.

Gabrys, Jennifer. "The Becoming Environmental of Computation: From Citizen Sensing to Planetary Computerization." *Tecnoscienza* 8, no. 1 (2017): 5–21.

Gailey, Phil. "Matching Congressmen and Executives, for a Price." *New York Times*, 5 May 1983.

Galler, Sidney R., and Basil R. Littin. "Economic Impact: Perspectives for Corporate Decision-Making." *Public Relations Journal* (1973): 10–12; 33.

Gamarekian, Barbara. "Foreign Image-Making: It's a Job for the Experts." *New York Times*, 11 October 1984.

Ganz, Marshall. *Why David Sometimes Wins: Leadership, Organization, and Strategy in the California Farm Worker Movement*. New York: Oxford, 2009.

Garcia-Johnson, Ronie. *Exporting Environmentalism: U.S. Multinational Chemical Corporations in Brazil and Mexico*. Cambridge: MIT Press, 2000.

Garey, Diane, and Lawrence R. Hott (dirs.). *The Wilderness Idea: John Muir, Gifford Pinchot, and the First Great Battle for Wilderness* [film]. Florentine Film, 1989.

Gelbspan, Ross. *Boiling Point: How Politicians, Big Oil and Coal, Journalists, and Activists Have Fueled a Climate Crisis—And What We Can Do to Avert Disaster*. New York: Basic Books, 2004.

Gelbspan, Ross. *The Heat Is On: The High Stakes Battle over Earth's Threatened Climate*. Reading, MA: Addison-Wesley, 1997.

Gismondi, Mike, and Debra Davidson. "Imagining the Tar Sands 1880–1967 and Beyond." *Imaginations* 3, no. 2 (2012): 68–103.

Gitelman, Howard M. *Legacy of the Ludlow Massacre: A Chapter in American Industrial Relations*. Philadelphia: University of Pennsylvania Press, 1988.

Global Environmental Management Initiative (GEMI). *Total Quality Environmental Management: The Primer*. Washington, DC: GEMI, 1993.

Global Environmental Management Initiative (GEMI). "Value to Business: Global Environmental Management Initiative." Washington, DC: GEMI, November 1998.

"Globe EU 1989–1999: Ten Years of Action for the Environment," n.d. http://www.globeinternational.org/publications/gi/rp01891299.pdf

GLOBE International. "History," 2017. https://globelegislators.org/.

"Golden Interview with John Hill." New York: Hill and Knowlton, 1975. http://industrydocuments.library.ucsf.edu/tobacco/docs/xyln0042.

Gonzalez, George A. *Corporate Power and the Environment: The Political Economy of U.S. Environmental Policy.* Lanham, MD: Rowman & Littlefield, 2001.

Gonzalez, George A. "The Conservation Policy Network, 1890–1910: The Development and Implementation of 'Practical' Forestry." *Polity* 31, no. 2 (1998): 269–299.

Goodwin, Doris Kearns. *The Bully Pulpit: Theodore Roosevelt, William Howard Taft, and the Golden Age of Journalism.* New York: Simon & Schuster, 2013.

Gordon, David J., and Craig A. Johnson. "The Orchestration of Global Urban Climate Governance: Conducting Power in the Post-Paris Climate Regime." *Environmental Politics* 26, no. 4 (2017): 694–714.

Graves, Henry. "A Policy of Forestry for the Nation." *Journal of Forestry* 17, no. 8 (1919): 901–910.

Green, Maia. "Calculating Compassion: Accounting for Some Categorical Practices in International Development." In *Adventures in Aidland: The Anthropology of Professionals in International Development,* ed. David Mosse, 33–47. New York: Berghahn Books, 2011.

Grubin, David (dir.). *The Image Makers* [film]. PBS, 1983. https://billmoyers.com/content/image-makers/.

Grunig, James E. "Review of Research on Environmental Public Relations." *Public Relations Review* 3, no. 3 (1977): 36–58.

Grunig, James E. "Communication Behaviors and Attitudes of Environmental Publics: Two Studies." *Journalism Monograph* 81 (Association for Education in Journalism and Mass Communication) (1983): 1–47.

Grunig, James E. "A Situational Theory of Environmental Issues, Publics, and Activists." In *Environmental Activism Revisited: The Changing Nature of Communication through Organizational Public Relations, Special Interest Groups and the Mass Media,* ed. Larissa Grunig. Monographs in Environmental Education and Environmental Studies, 5 (1989): 50–82.

Grunig, James E., ed. *Excellence in Public Relations and Communication Management.* Hillsdale, NJ: Lawrence Erlbaum, 1992.

Haas, Peter M. "Do Regimes Matter? Epistemic Communities and Mediterranean Pollution Control." *International Organization* 43, no. 3 (1989): 377–403.

Haas, Peter M. *Epistemic Communities, Constructivism, and International Environmental Politics.* New York: Routledge, 2016.

Habermas, Jürgen. *The Structural Transformation of the Public Sphere: An Inquiry into a Category of Bourgeois Society.* Cambridge: MIT Press, 1991.

Hahn, Tobias, Frank Figge, J. Alberto Aragón-Correa, and Sanjay Sharma. "Advancing Research on Corporate Sustainability: Off to Pastures New or Back to the Roots?" *Business & Society* 56, no. 2 (2017): 155–185.

Haines, Herbert. "Black Radicalization and the Funding of Civil Rights, 1957–1970." *Social Problems* 32, no. 1 (October 1984): 31–43.

Hajer, Maarten A. "'Verinnerlijking': The Limits to a Positive Management Approach." In *Environmental Law and Ecological Responsibility,* ed. Gunther Teubner, Lindsay Farmer, and Declan Murphy. New York: Wiley, 1994.

Hajer, Maarten, Måns Nilsson, Kate Raworth, Peter Bakker, Frans Berkhout, Yvo de Boer, Johan Rockström, Kathrin Ludwig, and Marcel Kok. "Beyond Cockpit-ism: Four

Insights to Enhance the Transformative Potential of the Sustainable Development Goals." *Sustainability* 7 (2015): 1651–1660.

Hallahan, Kirk. "Ivy Lee and the Rockefellers' Response to the 1913–1914 Colorado Coal Strike." *Journal of Public Relations Research* 14, no. 4 (2002): 265–315.

Hallahan, Kirk. "W. L. Mackenzie King: Rockefeller's 'Other' Public Relations Counselor in Colorado." *Public Relations Review* 29, no. 4 (2003): 401–414.

Harrison, E. Bruce. "A Quality Approach to Environmental Communication." *Environmental Quality Management* 1, no. 2 (1992): 225–231.

Harrison, E. Bruce. "Clean Air Act." *Public Relations Journal* (May 1978): 2.

Harrison, E. Bruce. "Counseling Companies on Environmental Communication." Washington, DC: EnviroComm International, 1995.

Harrison, E. Bruce. "Environment Energy: Public Relations at Large." *Public Relations Journal* 33, no. 2 (1977): 31.

Harrison, E. Bruce. "Environmental Health Committee Meeting." New York: Manufacturing Chemists' Association, 1966.

Harrison, E. Bruce. "EPA Reaches Out." *Public Relations Journal* (October 1978): 39.

Harrison, E. Bruce. "Grassroots Public Relations: The Art of Advocacy Stimulation to Affect Public Policy." Washington, DC: E. Bruce Harrison Company, 1987.

Harrison, E. Bruce. *Going Green: How to Communicate Your Company's Environmental Commitment.* Burr Ridge, IL: Business One Irwin, 1993.

Harrison, E. Bruce. "Green Communication in the Age of Sustainable Development: Gold Paper No. 9." International Public Relations Association, 1993.

Harrison, E. Bruce. "Management Guidelines/Clean Air Act '77: Part 2. Is 'No Growth' Really Ahead?" *Hydrocarbon Processing* 57, no. 8 (1978).

Harrison, E. Bruce. "Rule of Reason." *Public Relations Journal* (July 1978): 1–2.

Harrison, E. Bruce. "The Strategic Implications of Global Environmental Communication Needs." *Corporate Environmental Strategy* 3, no. 3 (1996): 77–83.

Harrison, E. Bruce. "Washington Focus." *Public Relations Journal* (1977): 9–10.

Hayden, F. Gregory, Alyx Dodds Garner, and Jerry Hoffman. "Corporate, Social, and Political Networks of Koch Industries Inc. and TD Ameritrade Holding Corporation: Extension to the State of Nebraska." *Journal of Economic Issues* 47, no. 1 (2013): 63–94.

Hays, Samuel P. *Conservation and the Gospel of Efficiency: The Progressive Conservation Movement, 1890–1920.* Cambridge, MA: Harvard University Press, 1959.

Heberlein, Thomas A. *Navigating Environmental Attitudes.* New York: Oxford University Press, 2012.

Hecox, Walter E. "Limits to Growth Revisited: Has the World Modeling Debate Made Any Progress?" *Boston College Environmental Affairs Law Review* 5, no. 1 (1976): 65–96. http://lawdigitalcommons.bc.edu/ealr/vol5/iss1/8.

Henderson, George L. *California & the Fictions of Capital.* New York: Oxford University Press, 1999.

Hertsgaard, Mark, and Kyle Pope. "Fixing the Media's Climate Failure." *The Nation* 308, no. 13 (May 2019): 12–21.

Hiebert, Ray Eldon. *Courtier to the Crowd: The Story of Ivy Lee and the Development of Public Relations.* Ames: Iowa State University Press, 1966.

Hill and Knowlton Inc. "Slings & Arrows, Inc: A Report on the Activists," 1971. https://www.industrydocuments.ucsf.edu/docs/#id=gpyd0051.

Hill, John W. *The Making of a Public Relations Man.* New York: D. McKay, 1963.

Hill, John W. "What We Learned from the Steel Negotiations." *Public Relations Journal* 16 (1960): 6–10.

Hill, John W., and Robert Skidelsky. "The Business of Business . . . The Government of Business." *New York Times*, 6 October 1976.

Hoffman, Andrew J. *From Heresy to Dogma: An Institutional History of Corporate Environmentalism.* Stanford, CA: Stanford University Press, 2001.

Hoggan, James, and Richard Littlemore. *Climate Cover-Up: The Crusade to Deny Global Warming.* Vancouver: Greystone Books, 2009.

Hoggan, James, and Marshall Ganz. "Sometimes David Wins." In *I'm Right and You're an Idiot: The Toxic State of Public Discourse and How to Clean It Up*, ed. James Hoggan, 173–186. Gabriola Island, BC: New Society Publishers, 2016.

Hosmer, Ralph S. "The Society of American Foresters: An Historical Summary." *Journal of Forestry* 48, no. 11 (1950): 756–777.

Hounshell, David A., and John K. Smith. *Science and Corporate Strategy: Du Pont R&D, 1902–1980.* Cambridge: Cambridge University Press, 1988.

Hull, John E. "Accomplishments in Air Pollution Control by the Chemical Industry." In *Proceedings of the National Conference on Air Pollution*, 61–64. Washington, DC: US Government Printing Office, 1958.

Igo, Sarah E. *The Averaged American: Surveys, Citizens, and the Making of a Mass Public.* Cambridge, MA: Harvard University Press, 2007.

Isin, Engin, and Evelyn Ruppert. "Data's Empire: Postcolonial Data Politics." In *Data Politics: Worlds, Subjects, Rights*, ed. Didier Bigo, Engin Isin, and Evelyn Ruppert, 208–227. London: Routledge, 2019.

IPRA. "Global Perspective on Environmental Communication Needs: Interim Report by the Environment Committee of the International Public Relations Association," Geneva: IPRA, 1995.

IPRA. *View from the Gallery: A News Bulletin of the International Public Relations Association*, ed. Pierre-André Hervo. Geneva: IPRA, 1993.

Iyengar, Shanto, and Douglas S. Massey. "Scientific Communication in a Post-Truth Society." *Proceedings of the National Academy of Sciences of the United States of America* 116, no. 16 (2019): 7656–7661.

Jackson, Brooks. "Easy Money: U.S. Lawmakers' Take from Honorariums Hits $10 Million a Year—It's All Legal, but Payments by Lobbies Reach a Level Some View as Scandalous—Case of the Naive Prosecutor." *Wall Street Journal*, 1 November 1988.

Jansen, Sue Curry. "Semantic Tyranny: How Edward L. Bernays Stole Walter Lippmann's Mojo and Got Away with It and Why It Still Matters." *International Journal of Communication* 7 (2013): 1094–1111.

Jarvik, Laurence. "PBS and the Politics of Quality: Mobil Oil's "Masterpiece Theatre."" *Historical Journal of Film, Radio and Television* 12, no. 3 (1992): 253–274.

Jasanoff, Sheila. "Procedural Choices in Regulatory Science." *Technology in Society* 17, no. 3 (1995): 279–293.

Jasanoff, Sheila. *The Fifth Branch: Science Advisers as Policymakers.* Cambridge, MA: Harvard University Press, 1990.

John, Steve, and Stuart Thomson, eds. *New Activism and the Corporate Response.* New York: Palgrave Macmillan, 2003.

"John W. Hill, 86, Dies; Led Hill & Knowlton." *New York Times*, 18 March 1977.

Johnson, Dennis W. *Democracy for Hire: A History of American Political Consulting.* New York: Oxford University Press, 2017.

Johnson, Robert Underwood. "A High Price to Pay for Water: Apropos of the Grant of the Hetch-Hetchy Valley to San Francisco for a Reservoir." *Century Magazine* 76 (1908): 632–634. https://babel.hathitrust.org/cgi/pt?id=inu.32000000493215&view=1up&seq=13.

Johnson, Robert Underwood. *Remembered Yesterdays*. Boston: Little, Brown, 1923.

Johnson, Robert Underwood. "The Yosemite National Park." *Outlook*, February 1909. https://babel.hathitrust.org/cgi/pt?id=iau.31858033603295&view=1up&seq=7.

Jones, Barrie L., and W. Howard Chase. "Managing Public Policy Issues." *Public Relations Review* 5, no. 2 (1979): 3–23.

"Juice: The Future of Power and Influence in Washington." *Inside PR*, May 1992.

"Kenneth Bousquet Dies, Former Senate Counsel." *Washington Post*, 9 October 1977.

Kenworthy, Tom. "Courting the Key Committees; Industry Honoraria Flow to Those with Jurisdiction, Analysis Finds." *Washington Post*, 3 August 1988.

Kerr, Robert L. *The Rights of Corporate Speech: Mobil Oil and the Legal Development of the Voice of Big Business*. New York: LFB Scholarly Publishing, 2005.

Kirsch, Stuart. *Mining Capitalism: The Relationship Between Corporations and Their Critics*. Oakland: University of California Press, 2014.

Knorr Cetina, Karin. *Epistemic Cultures: How the Sciences Make Knowledge*. Cambridge, MA: Harvard University Press, 1999.

Krause, Monika. *The Good Project: Humanitarian Relief NGOs and the Fragmentation of Reason*. Chicago: University of Chicago Press, 2014.

Kuethe, Rik. "Access as Bargaining Chip (Toegang Als Pasmunt)." *Elsevier Weekly Magazine* (in Dutch), 1989.

Kuneva, Meglena. 2009. Keynote Speech: Roundtable on Online Data Collection, Targeting and Profiling. Speech 09/156. European Commission, 31 March 2009. https://ec.europa.eu/commission/presscorner/detail/en/SPEECH_09_156.

Laidler, John. "High Tech Is Watching You." *Harvard Gazette*, March 2019.

Lambert, Kate. "Scenic Hudson and Storm King: Revolutionizing Standing in Environmental Litigation." *Michigan Journal of Environmental & Administrative Law*, 1 December 2014.

Latour, Bruno. *An Inquiry into Modes of Existence: An Anthropology of the Moderns*. Cambridge, MA: Harvard University Press, 2013.

Latour, Bruno. "From Realpolitik to Dingpolitik or How to Make Things Public." In *Making Things Public: Atmospheres of Democracy*, ed. Bruno Latour and Peter Weibel, 14–41. Cambridge, MA: MIT Press, 2005.

Latour, Bruno. *Pandora's Hope: Essays on the Reality of Science Studies*. Cambridge, MA: Harvard University Press, 1999.

Laumann, Edward O., and David Knoke. *The Organizational State: Social Choice in National Policy Domains*. Madison: University of Wisconsin Press, 1987.

Lazarus, Richard. "Super-wicked Problems and Climate Change: Restraining the Present to Liberate the Future." *Cornell Law Review* 94 (2009): 1153–1234.

Lee, Caroline W., Michael McQuarrie, and Edward T. Walker. *Democratizing Inequalities: Dilemmas of the New Public Participation*. New York: New York University Press, 2015.

Lee, Ivy. "Enemies of Publicity." *Electric Railway Journal* 49, no. 13 (1917): 599–600.

Lee, Ivy. *Publicity: Some of the Things It Is and Is Not*. New York: Industries Publishing, 1925.

LeMenager, Stephanie. *Living Oil: Petroleum Culture in the American Century*. Oxford: Oxford University Press, 2014.

Lerbinger, Otto. "A Long View of the Environment." *Public Relations Journal* 29 (1973): 20–21.

Lesly, Philip. "Survival in an Age of Activism." *Public Relations Journal* 25 (1969): 6–8.

Letouzé, Emmanuel, and Patrick Vinck. "A New and Sometimes Awkward Relationship." Data Pop Alliance, 29 January 2015. https://datapopalliance.org/big-data-and-human-rights-a-new-and-sometimes-awkward-relationship-to-be-further-explored/.

Levick, Richard S. "The Interview (Richard S. Levick)." *The Native Influence*, n.d. https://natfluence.com/interview/rslevick/#interview.

Levy, David L. "Environmental Management as Political Sustainability." *Organization & Environment* 10, no. 2 (1997): 126–147.

Levy, David L., and Peter J. Newell, eds. *The Business of Global Environmental Governance.* Cambridge, MA: MIT Press, 2004.

Libbey, Mary Beth. "Conservation and the Corporation." *Across the Board: The Conference Board Magazine* 15, no. 4 (April 1978).

Lippert, Ingmar. "Failing the Market, Failing Deliberative Democracy: How Scaling Up Corporate Carbon Reporting Proliferates Information Asymmetries." *Big Data & Society* 3, no. 2 (2016): 1–13.

Lippert, Ingmar. "Corporate Carbon Footprinting as Techno-Political Practice." In *The Carbon Fix: Forest Carbon, Social Justice, and Environmental Governance*, ed. Stephanie Paladino and Shirley J. Fiske, 197–118. London: Routledge 2016.

Lippmann, Walter. *A Preface to Politics.* New York: Mitchell Kennerley, 1913.

Lippmann, Walter. *Public Opinion.* New York: Simon and Schuster, [1922] 1997.

Lloyd, H. D. "The Story of a Great Monopoly." *The Atlantic* (Boston, March 1881). https://www.theatlantic.com/magazine/archive/1881/03/the-story-of-a-great-monopoly/306019/.

Lohmann, Larry. "Marketing and Making Carbon Dumps: Commodification, Calculation and Counterfactuals in Climate Change Mitigation." *Science as Culture* 14, no. 3 (2005): 203–235.

Lounsbury, Michael, Marc Ventresca, and Paul M. Hirsch. "Social Movements, Field Frames and Industry Emergence: A Cultural-Political Perspective on US Recycling." *Socio-Economic Review* 1, no. 1 (2003): 71–104.

MacDonald, Andrew H. "National Harbor a Threat to the Potomac." *Baltimore Sun*, 7 January 1998.

MacLowry, Randall (dir.). *The Mine Wars* [film]. PBS, 2016.

Mah, Alice. "Environmental Justice in the Age of Big Data: Challenging Toxic Blind Spots of Voice, Speed, and Expertise." *Environmental Sociology* 3, no. 2 (2017): 122–133.

Major, Ann Marie. "Environmental Concern and Situational Communication Theory." *Journal of Public Relations Research* 5, no. 4 (2009): 251–268.

Manheim, Jarol B. *Strategy in Information and Influence Campaigns: How Policy Advocates, Social Movements, Insurgent Groups, Corporations, Governments, and Others Get What They Want*, New York: Routledge, 2011.

Marchand, Roland. *Creating the Corporate Soul: The Rise of Public Relations and Corporate Imagery in American Big Business*, Berkeley: University of California Press, 1998.

Marr, Bernard. "A Brief History of Big Data Everyone Should Read." WEF Blog, 25 February 2015. https://www.weforum.org/agenda/2015/02/a-brief-history-of-big-data-everyone-should-read/.

Marres, Noortje. "Issues Spark a Public into Being: A Key but Often Forgotten Point of the Lippmann-Dewey Debate." In *Making Things Public: Atmospheres of Democracy*, ed. Bruno Latour and Peter Weibel, 208–217. Cambridge, MA: MIT Press, 2005.

Marres, Noortje. "The Issues Deserve More Credit: Pragmatist Contributions to the Study of Public Involvement in Controversy." *Social Studies of Science* 37, no. 5 (2007): 759–780.

Martin, Isaac William. *Rich People's Movements*. New York: Oxford University Press, 2013.

Matz, Jacob, and Daniel Renfrew. "Selling 'Fracking': Energy in Depth and the Marcellus Shale." *Environmental Communication* 9, no. 3 (2015): 288–306.

Mayer, Jane. *Dark Money: The Hidden History of the Billionaires Behind the Rise of the Radical Right*. New York: Doubleday, 2016.

McCraw, Thomas K. "Business & Government: The Origins of the Adversary Relationship." *California Management Review* 26 no. 2 (Winter 1984): 33–52.

McCright, Aaron M. "Anti-Reflexivity and Climate Change Skepticism in the US General Public." *Human Ecology Review* 22, no. 2 (2016): 77–108.

McCright, Aaron M., and Riley E. Dunlap. "Anti-Reflexivity: The American Conservative Movement's Success in Undermining Climate Science and Policy." *Theory, Culture and Society* 27, no. 2 (2010): 100–133.

McCright, Aaron M., and Riley E. Dunlap. "The Politicization of Climate Change and Polarization in the American Public's Views of Global Warming, 2001–2010." *Sociological Quarterly* 52, no. 2 (2016): 155–194.

McFarland, Andrew S. *Cooperative Pluralism: The National Coal Policy Experiment*. Lawrence: University Press of Kansas, 1993.

McGeary, Martin Nelson. *Gifford Pinchot: Forester-Politician*. Princeton, NJ: Princeton University Press, 1960.

McGoey, Linsey. *No Such Thing as a Free Gift: The Gates Foundation and the Price of Philanthropy*. New York: Verso, 2015.

McGoey, Linsey. "Philanthrocapitalism and Its Critics." *Poetics* 40, no. 2 (2012): 185–199.

McKeever, Amy. "Trump's and Biden's Environmental Policy Promises and Actions." *National Geographic*, Washington, D.C., September 2020. https://www.nationalgeographic.com/science/trackers/latest-trump-biden-environmental-policy-promises-actions/.

Meadows, Donella H., Dennis L. Meadows, Jorgen Randers, and William W. Behrens. *The Limits to Growth: A Report for the Club of Rome's Project on the Predicament of Mankind*. New York: Universe Books, 1972.

Melnick, R. Shep. *Regulation and the Courts: The Case of the Clean Air Act*. Washington, DC: Brookings Institution, 1983.

Meuser, Michael, and Ulrike Nagel. "The Expert Interview and Changes in Knowledge Production." In *Interviewing Experts*, ed. Alexander Bogner, Beate Littig, and Wolfgang Menz, 17–42. London: Palgrave Macmillan, 2009.

Meyer, John M. "Gifford Pinchot, John Muir, and the Boundaries of Politics in American Thought." *Polity* 30, no. 2 (1997): 267–284.

Michel, Dominique. "A Bruxelles Si Vous Ne Voulez Pas La Subir." *L'Entreprise*, 1990. https://www.industrydocuments.ucsf.edu/docs/#id=yfhm0201

Miles, Riley S. "Maintaining an Environmental Balance." *Environmental Science & Technology* 10, no. 5 (1976): 418–419.

Miller, Char. *Gifford Pinchot and the Making of Modern Environmentalism*. Washington, DC: Island Press, 2001.

Miller, David, and William Dinan. *A Century of Spin: How Public Relations Became the Cutting Edge of Corporate Power*. London: Pluto Press, 2007.

Miller, Karen S. *The Voice of Business: Hill & Knowlton and Postwar Public Relations.* Chapel Hill: University of North Carolina Press, 1999.

Mitchell, Timothy. *Carbon Democracy: Political Power in the Age of Oil.* London: Verso, 2013.

Mitchell, Timothy. *Rule of Experts: Egypt, Techno-Politics, Modernity.* Berkeley: University of California Press, 2002.

Mitchell, Robert Cameron, Angela G. Mertig, and Riley E. Dunlap. "Twenty Years of Environmental Mobilization: Trends Among National Environmental Organizations." *Society & Natural Resources* 4 no. 3 (1991): 219–234.

Mizruchi, Mark S. *The Structure of Corporate Political Action: Interfirm Relations and Their Consequences.* Cambridge, MA: Harvard University Press, 1992.

Mobil Oil. "Evolution of Mobil's Public Affairs Programs 1970–81." Fairfax, VA: Mobil Oil, 1982.

Moore, Richard L. "Environment—A New PR Crisis." *Public Relations Journal* 26 (1970): 6–9.

Moore, Susan. "Environmental Improvement through Business Incentives." Washington, DC: GEMI, 1999.

Moore, W. J. "Have Smarts, Will Travel." *National Journal,* 28 November 1987: 3020–3025.

Morris, Edmund. *Theodore Rex.* New York: Random House, 2001.

Mosse, David. *Cultivating Development: An Ethnography of Aid Policy and Practice.* Ann Arbor, MI: Pluto Press, 2005.

Mosse, David. "Global Governance and the Ethnography of International Aid." In *The Aid Effect: Giving and Governing in International Development,* ed. David Mosse and David Lewis, 1–36. Ann Arbor, MI: Pluto Press, 2005.

Moyer, Reed. "Where We Agree—A Report of the National Coal Policy Project." *Natural Resources Journal* 18, no. 4 (1978): 969–971.

Muir, John. "Features of the Proposed Yosemite National Park." *Century Magazine* 40, no. 5 (1890): 656–667.

Muir, John. "The American Forests." *Atlantic Monthly* 80, no. 478 (1897): 145–157.

Muir, John. "The National Parks and Forest Reservations." *Harper's Weekly* 41, no. 2111 (1897): 563–567.

Muir, John. "The Treasures of the Yosemite." *Century Magazine* 40, no. 4 (1890): 483–500.

Muir, John. "The Wild Parks and Forest Reservations of the West." *The Atlantic* (Boston, January 1898). https://www.theatlantic.com/magazine/archive/1898/01/the-wild-parks-and-forest-reservations-of-the-west/544038/.

Muir, John, and Marion Randall Parsons. "Preface." In *Travels in Alaska,* 5–6. Boston: Houghton Mifflin, 1915.

Mumford, John Kimberly. "This Land of Opportunity: The Attitude of Great Corporations Towards Their Men." *Harper's Weekly* 52, no. 1 (1908): 20–23, 29.

Munkirs, John R., and James I. Sturgeon. "Oligopolistic Cooperation: Conceptual and Empirical Evidence of Market Structure Evolution." *Journal of Economic Issues* 19, no. 4 (1985): 899–921.

Munshi, Debashish, and Lee Edwards. "Understanding 'Race' in/and Public Relations: Where Do We Start and Where Should We Go?" *Journal of Public Relations Research* 23, no. 4 (2011): 349–367.

Murphy, Priscilla Coit. *What a Book Can Do: The Publication and Reception of Silent Spring.* Amherst: University of Massachusetts Press, 2005.

"Nairobi Code for Communication on Environment and Development." *Environmental Conservation* 20, no. 1 (1993): 85.

Nash, Roderick Frazier. *Wilderness and the American Mind*. 5th ed. New Haven, CT: Yale University Press, 2014.

"National Coal Policy Project a Mixed Success." *Chemical & Engineering News*, March 1980.

National Geographic. "A Running List of How President Trump Is Changing Environmental Policy." *National Geographic*, May 2019.

National Labor Relations Board. "1935 Passage of the Wagner Act." https://www.nlrb.gov/about-nlrb/who-we-are/our-history/1935-passage-of-the-wagner-act.

NEDA. "Invitation to the National Environmental Development Association (NEDA) Conference on Regulatory Issues," National Environmental Development Association, 1979. https://ufdc.ufl.edu/WL00002412/00001/print?options=1JJ&options=1JJ.

"New Hope for the Forests: Encouraging Responses from Governors to the Proposal of a Conference." *Century Magazine* 75 (1908): 634–638. https://babel.hathitrust.org/cgi/pt?id=inu.32000000493223&view=1up&seq=658.

"New Ways to Lobby a Recalcitrant Congress." *Business Week*, 3 September 1979.

"Next Steps in Forestry Reform." *Century Magazine* 69 (1904): 315–316. https://babel.hathitrust.org/cgi/pt?id=mdp.39076002641152&view=1up&seq=329.

Olasky, Marvin N. *Corporate Public Relations: A New Historical Perspective*. Hillsdale, NJ: L. Erlbaum, 1987.

Oravec, Christine. "Conservationism vs. Preservationism: The 'Public Interest' in the Hetch Hetchy Controversy." *Quarterly Journal of Speech* 70, no. 4 (1984): 444–458.

Oreskes, Naomi, and Erik M. Conway. *Merchants of Doubt: How a Handful of Scientists Obscured the Truth on Issues from Tobacco Smoke to Global Warming*. New York: Bloomsbury Press, 2010.

Otterbourg, Robert K. "Public Relations Pioneers . . . Gifford Pinchot: Conservationist and Publicist." *Public Relations Quarterly* 19, no. 1 (1974): 19–23.

Parenti, Christian. "'The Limits to Growth': A Book That Launched a Movement." *The Nation*, December 2012.

Parisi, Anthony J. "Book Brings the Rule of Reason to Corporation-Public Clashes." *New York Times*, 10 February 1978.

Parks, Lisa. "Around the Antenna Tree: The Politics of Infrastructural Visibility." *Flow: A Critical Forum on Media & Culture* (6 March 2009): 345–347.

Pasek, Anne. "Managing Carbon and Data Flows: Fungible Forms of Mediation in the Cloud." *Culture Machine* (2019): 1–15.

Pasek, Anne. "Mediating Climate, Mediating Scale." *Humanities* 8 no. 4 (2019): 1–13.

Pattberg, Philipp H. *Private Institutions and Global Governance: The New Politics of Environmental Sustainability*. Northampton, MA: Edward Elgar, 2007.

Peters, John Durham. "Democracy and American Mass-Communication Theory: Dewey, Lippmann, Lazarsfeld." *Communication* 11, no. 3 (1989): 199–220.

Peterson Companies. "Waterfront Development Exceeds 100 Sales in 12-Month Period." National Harbor, MD: National Harbor, 2011. https://www.nationalharbor.com/press-releases/the-peterson-companies-national-harbor-among-leaders-in-residential-sales-volume-throughout-the-dc-metro-area/.

Phelan, James D. "Why Congress Should Pass the Hetch-Hetchy Bill." *Outlook* 91, no. 7 (13 February 1909): 340.

Pierini, Bruce. "How Did the Hetch Hetchy Project Impact Native Americans?" *Sierra College: Snowy Range Reflections* 6, no. 1 (2015). https://www.sierracollege.edu/ejournals/jsnhb/v6n1/pierini.html.

Pinchot, Gifford. *Breaking New Ground*. New York: Harcourt, Brace and Company, 1947.

Pinchot, Gifford. "Bulletin to the Members of the National Conservation Association." Washington, DC: National Conservation Association, 30 June 1916. https://babel. hathitrust.org/cgi/pt?id=umn.31951p00713425d&view=1up&seq=1.

Pinchot, Gifford. *Gifford Pinchot: Selected Writings*. Edited by Char Miller. University Park: Pennsylvania State University Press, 2017.

Pinchot, Gifford. *The Fight for Conservation*. New York: Doubleday, Page, 1910. https:// babel.hathitrust.org/cgi/pt?id=uc2.ark:/13960/t1ng4t61d&view=1up&seq=7.

Pinchot, Gifford. "The Lines Are Drawn." *Journal of Forestry* 17, no. 8 (1919): 899–900.

Pinchot, Gifford. "The Profession of Forestry." Washington, DC: American Forestry Association, 1901. https://foresthistory.org/wp-content/uploads/2017/01/Pinchot_Forestry_.pdf.

Pinchot, Gifford. "The Use of the National Forests." Washington, DC: USDA Forest Service, 1907.

Pinchot, Gifford. "Where We Stand." *Journal of Forestry* 18, no. 5 (1920): 441–447.

Pinkett, Harold T. "Gifford Pinchot, Consulting Forester, 1893–1898." *New York History* 39, no. 1 (1958): 34–49.

Pinkett, Harold T. *Gifford Pinchot: Private and Public Forester*. Urbana: University of Illinois Press, 1970.

Pinkett, Harold T. "The Forest Service, Trail Blazer in Recordkeeping Methods." *American Archivist* 22, no. 4 (1959): 419–426.

Ponder, Stephen. "Federal News Management in the Progressive Era: Gifford Pinchot and the Conservation Crusade." *Journalism History* 13, no. 2 (1986): 42–48.

Ponder, Stephen. "Gifford Pinchot: Press Agent for Forestry." *Journal of Forest History* 31, no. 1 (1987): 26–35.

Ponder, Stephen. "Progressive Drive to Shape Public Opinion, 1898–1913." *Public Relations Review* 16, no. 3 (1990): 94–104.

Pooley, Eric. *The Climate War: True Believers, Power Brokers, and the Fight to Save the Earth*. New York: Hyperion, 2010.

Poom, Age, Olle Järv, Matthew Zook, and Tuuli Toivonen. "COVID-19 Is Spatial: Ensuring that Mobile Big Data Is Used for Social Good." *Big Data & Society*, July-December 2020: 1–7.

Poovey, Mary. *Genres of the Credit Economy Mediating Value in Eighteenth- and Nineteenth-Century Britain*. Chicago: University of Chicago Press, 2008.

Popoff, Frank. "Corporate America: An Agenda for What's Right in the '90s." Public Relations Society of America National Conference, Phoenix, AZ, PRSA, Box 171–3, 4 November 1991.

Porter, Michael E., and Mark R. Kramer. "Creating Shared Value." *Harvard Business Review* 89 (January-February 2011): 62–77.

Porter, Michael E., and Mark R. Kramer. "Strategy and Society: The Link Between Competitive Advantage and Corporate Social Responsibility." *Harvard Business Review* 84, no. 12 (December 2006): 78–92.

Powell, Lewis F. "Attack on American Free Enterprise System." Lewis F. Powell, Jr., Archives, 1971. https://lawdigitalcommons.bc.edu/cgi/viewcontent.cgi?article=1078 &context=darter_materials.

Power, Michael. *Organized Uncertainty: Designing a World of Risk Management*. Oxford: Oxford University Press, 2007.

Power, Michael. "Expertise and the Construction of Relevance: Accountants and Environmental Audit." *Accounting, Organizations and Society* 22, no. 2 (1997): 123–146.

Pulver, Simone. "Making Sense of Corporate Environmentalism: An Environmental Contestation Approach to Analyzing the Causes and Consequences of the Climate Change Policy Split in the Oil Industry." *Organization and Environment* 20, no. 1 (2007): 44–83.

Quarles, John. "A Thicket of Environmental Laws." *Wall Street Journal*, 24 August 1979.

Quarles, John. "EMB: Congress at Its Worst." *Wall Street Journal*, 13 November 1979.

Quarles, John. "Maturing Environmentalism." *Journal of the Air Pollution Control Association* 31, no. 9 (1981): 967–969.

Quarles, John. "The Clean Air Amendments." *Wall Street Journal*, 28 December 1977.

Rabin-Havt, Ari. *Lies, Incorporated: The World of Post-Truth Politics*. New York: Anchor Books, 2016.

Rajpurohit, Anmol. "Interview: Emanuel Letouzé, Data-Pop Alliance on Big Data and Human Rights." KD Nuggets, 2015. https://www.kdnuggets.com/2015/04/interview-emmanuel-letouze-data-pop-alliance-human-rights.html.

Regalzi, Francesco. "Democracy and Its Discontents: Walter Lippmann and the Crisis of Politics (1919–1938)." *E-Rea* 9 (2012): 1–9.

Revzin, Philip. "Brussels Babel: European Bureaucrats Are Writing the Rules Americans Will Live By." *Wall Street Journal*, 17 May 1989.

Revzin, Philip. "World Business (A Special Report): The Uncommon Market—United We Stand . . . as Europe Moves Toward Unity in 1992, the Brussels Bureaucrat Is in the Driver's Seat." *Wall Street Journal*, 22 September 1989.

Rich, Laurie A., and Kenneth Jacobson. "Alternative Dispute Resolution—Opening Doors to Settlements." *Chemical Week*, 14 August 1985.

Righter, Robert W. *The Battle over Hetch Hetchy: America's Most Controversial Dam and the Birth of Modern Environmentalism*. New York: Oxford University Press, 2005.

Rockefeller, David. "Free Trade in Ideas." *Chief Executive Magazine* no. 6, Autumn 1978.

"Rockefeller Plies Pick in Coal Mine; Dons Overalls and Jumper and Makes First-Hand Investigation of Colorado Conditions. Calls Men His Partners. Tells Them Their Interests Are Similar; Questions Coal Diggers about Wages and Work." *New York Times*, 22 September 1915. https://timesmachine.nytimes.com/timesmachine/1915/09/22/104654529.html?pageNumber=5.

Rogers, Douglas. "The Materiality of the Corporation: Oil, Gas, and Corporate Social Technologies in the Remaking of a Russian Region." *American Ethnologist* 39, no. 2 (2012): 284–296.

Roosevelt, Theodore. *Theodore Roosevelt: An Autobiography*. New York: Macmillan, 1913.

Ross, Benjamin, and Steven Amter. *The Polluters: The Making of Our Chemically Altered Environment*. Oxford: Oxford University Press, 2010.

Ross, Brendan D. "From Practical Woodsman to Professional Forester: Henry S. Graves and the Professionalization of Forestry in the United States, 1900–1920." *MSSA Kaplan Prize for Use of MSSA Collections*. 2. https://elischolar.library.yale.edu/mssa_collections/2.

Rothwell, Jerry (dir.). *How to Change the World* [film]. Impact Partners, 2015.

Roy, William G., and Rachel Parker-Gwin. "How Many Logics of Collective Action?" *Theory and Society* 28, no. 2 (1999): 203–237.

Russel, Trevor. "Corporate Environmental Disclosure: Is Business Measuring Up to Its Responsibilities?" *Corporate Environmental Disclosure* 4, no. 2 (1995): 137–140.

Russell, Adrienne, and Matthew Tegelberg. "Beyond the Boundaries of Science: Resistance to Misinformation by Scientist Citizens." *Journalism* 21, no. 3 (2019): 327–344.

Russill, Chris. "Dewey/Lippmann Redux." *Empedocles: European Journal for the Philosophy of Communication* 7, no. 2 (2016): 129–142 .

Russill, Chris. "Through a Public Darkly: Reconstructing Pragmatist Perspectives in Communication Theory." *Communication Theory* 18, no. 4 (2008): 478–504.

Russill, Chris. "Looking for the Horizon: A Conversation Between John Durham Peters and Chris Russill. *Canadian Journal of Communication* 42 (2017): 683–699.

Russill, Chris. "Truth and Opinion in Climate Change Discourse: The Gore-Hansen Disagreement." *Public Understanding of Science* 20, no. 6 (2011): 796–809.

Salganik, Matthew J. *Bit by Bit: Social Research in the Digital Age.* Princeton, NJ: Princeton University Press, 2017.

Sallada, Logan H., and Brendan G. Doyle, eds. *The Spirit of Versailles: The Business of Environmental Management.* Paris: ICC, 1986.

Schlozman, Kay Lehman, and John T. Tierney. *Organized Interests and American Democracy.* New York: Harper & Row, 1986.

Schmertz, Herbert, and William Novak. *Good-Bye to the Low Profile: The Art of Creative Confrontation.* Boston: Little, Brown, 1986.

Schmidheiny, Stephan, Rodney Chase, and Livio DeSimone, "Signals of Change: Business Progress toward Sustainable Development." Geneva: WBCSD, 1997.

Schneider, Jen, Steve Schwarze, Peter K. Bsumek, and Jennifer Peeples. *Under Pressure: Coal Industry Rhetoric and Neoliberalism.* London: Palgrave Macmillan, 2016.

Schoklitsch, Hanno. "Climate Change and Big Data: Investing for a Solution." *Forbes,* 6 September 2019.

Schudson, Michael. *Discovering the News: A Social History of American Newspapers.* New York: Basic Books, 1978.

Schudson, Michael. "The 'Lippmann-Dewey Debate' and the Invention of Walter Lippmann as an Anti-Democrat 1986–1996." *International Journal of Communication* 2 (2008): 1031–1042.

Schudson, Michael. "Walter Lippmann's Ghost: An Interview with Michael Schudson." *Mass Communication and Society* 19, no. 3 (2016): 221–229.

Schudson, Michael. *The Rise of the Right to Know: Politics and the Culture of Transparency, 1945–1975.* Cambridge, MA: Harvard University Press, 2015.

Schuler, Douglas A. "Corporate Political Action: Rethinking the Economic and Organizational Influences." *Business and Politics* 1, no. 1 (1999): 83–97.

Scott, James C. *Seeing Like a State: How Certain Schemes to Improve the Human Condition Have Failed.* New Haven, CT: Yale University Press, 1998.

"Self-Evident Subtlety." *TIME Magazine,* 1 August 1938.

Sellers, Christopher C. *Hazards of the Job: From Industrial Disease to Environmental Health Science.* Chapel Hill: University of North Carolina Press, 1997.

Sethi, S. Prakash. "Corporate Political Activism." *California Management Review* 24, no. 3 (1982): 32–42.

Sethi, S. Prakash. "Serving the Public Interest: Corporate Political Action Strategies for the 1980s." *Management Review* 70, no. 3 (1981: 8–11).

Sethi, S. Prakash and Herbert Schmertz. "Industry Fights Back: The Debate over Advocacy Advertising." *Saturday Review,* 21 January 1978: 20-25. https://www.industrydocuments.ucsf.edu/docs/mywc0105.

Sewell, William H. "Historical Events as Transformations of Structures: Inventing Revolution at the Bastille." In *Logics of History: Social Theory and Social Transformation,* 225–270. Chicago: University of Chicago Press, 2005.

Shants, Frank B. "Countering the Anti-Nuclear Activists." *Public Relations Journal* 34, no. 10 (1978): 10.

Shared Value Initiative. "About Shared Value Initiative: Creating Unexpected Connections." 2020. https://www.sharedvalue.org/about/.

Sheingate, Adam. *Building a Business of Politics: The Rise of Political Consulting and the Transformation of American Democracy*. New York: Oxford University Press, 2016.

Shellenberger, Michael, and Ted Nordhaus. "The Death of Environmentalism: Global Warming Politics in a Post-Environmental World." *Geopolitics, History & International Relations* 1, no. 1 (2009): 121–163.

Sicilia, David B. "The Corporation Under Siege: Social Movements, Regulation, Public Relations, and Tort Law Since the Second World War." In *Constructing Corporate America: History, Politics, Culture*, ed. Kenneth Lipartito and David B. Sicilia. Oxford: Oxford University Press, 2004.

Silverstein, Ken. *The Secret World of Oil*. London: Verso, 2014.

Simon, Rita James. "Public Attitudes Toward Population and Pollution." *Public Opinion Quarterly* 35, no. 1 (1971): 93–98.

Sinclair, Upton Beall. *Oil*. New York: Washington Square Press, 1966.

Sklair, Leslie. "The Transnational Capitalist Class and the Discourse of Globalisation." *Cambridge Review of International Affairs* 14, no. 1 (2000): 67–85.

Smith, Herbert A. "The Early Forestry Movement in the United States." *Agricultural History* 12, no. 4 (1938): 326–346.

Smith, Philip, and Nicholas Howe. *Climate Change as Social Drama: Global Warming in the Public Sphere*. Cambridge: Cambridge University Press, 2016.

Sonnenfeld, Jeffrey A. *Corporate Views of the Public Interest: Perceptions of the Forest Products Industry*. Boston: Auburn House, 1981.

Spillman, Lyn. *Solidarity in Strategy: Making Business Meaningful in American Trade Associations*. Chicago: University of Chicago Press, 2012.

St. John, Burton III. "The 'Creative Confrontation' of Herbert Schmertz: Public Relations Sense Making and the Corporate Persona." *Public Relations Review* 40, no. 5 (2014): 772–779.

Stamm, Keith R. "Conservation Communications Frontiers: Reports of Behavioral Research." In *Interpreting Environmental Issues: Research and Development in Conservation Communications*, ed. Clay Schoenfeld. Madison, WI: Dembar Educational Research Services, 1973.

Stamm, Keith R. "Two Orientations to the Conservation Concept of Scarcity." *Environmental Education* 1, no. 4 (1970): 134–139.

Stamm, Keith R., and John E. Bowes. "Environmental Attitudes and Reaction." *Environmental Education* 3, no. 3 (1972): 56–60.

Stamm, Keith R., and James E. Grunig. "Communication Situations and Cognitive Strategies in Resolving Environmental Issues." *Journalism Quarterly* 54, no. 4 (Winter 1977): 713–720.

Stark, David. *The Sense of Dissonance: Accounts of Worth in Economic Life*. Princeton, NJ: Princeton University Press, 2009.

Stauber, John, and Sheldon Rampton. *Toxic Sludge Is Good for You!* Monroe, ME: Common Courage Press, 1995.

"Steel Company Pays $235,000 to Settle $4,643,000 in Donora Smog Death Suits." *New York Times*, 18 April 1951.

Steen, Harold K. *The U.S. Forest Service: A History*. Seattle: University of Washington Press, 1976.

Stone, Diane. "Transfer Agents and Global Networks in the 'Transnationalization' of Policy." *Journal of European Public Policy* 11, no. 3 (2004): 545–566.

Suchman, Mark C. "Managing Legitimacy: Strategic and Institutional Approaches." *Academy of Management Review* 20, no. 3 (1995): 571–610.

Sudman, Seymour, and Norman M. Bradburn. "The Organizational Growth of Public Opinion Research in the United States." *Public Opinion Quarterly* 51, Part 2 (1987): S67-S78.

Supran, Geoffrey, and Naomi Oreskes. "Assessing ExxonMobil's Climate Change Communications (1977–2014)." *Environmental Research Letters* 12, no. 8 (2017): 1–18.

Swetonic, Matthew M. "Death of the Asbestos Industry." In *Crisis Response: Inside Stories on Managing Image Under Siege*. Detroit: Gale Research, 1993. https://www.industrydocuments.ucsf.edu/docs/#id=nygy0089.

Tarbell, Ida. *The History of the Standard Oil Company*. New York: Macmillan, 1925.

Tatevossian, Anoush Rima. "Data Philanthropy: Public & Private Sector Data Sharing for Global Resilience." UN Global Pulse Blog, 2011. http://www.unglobalpulse.org/blog/data-philanthropy-public-private-sector-data-sharing-global-resilience.

Tatevossian, Anoush Rima. "Mapping the Next Frontier of Open Data: Corporate Data Sharing." UN Global Pulse Blog, 2014. https://www.unglobalpulse.org/2014/09/mapping-the-next-frontier-of-open-data-corporate-data-sharing/.

Tedlow, Richard S. *Keeping the Corporate Image: Public Relations and Business, 1900–1950*. Greenwich: JAI Press, 1979.

"The Amplifier (De Geluidsversterker)." *Trends Financial Magazine*, 9 November 1989: 66.

"The Need of a National Forest Commission." *Century Magazine* 49 (1895): 634–635. https://babel.hathitrust.org/cgi/pt?id=inu.32000000493488&view=1up&seq=7.

The Story of DDT [film]. Great Britain, Army Kinematograph Service, 1944. https://www.youtube.com/watch?v=1F8wTX-yidM.

Thompson, Carl. "Communicators and Their Environmental Problems." *Public Relations Journal* 29 (1973): 34–35.

Tichenor, P. J., G. A. Donohue, C. N. Olien, and J. K. Bowers. "Environment and Public Opinion." *Journal of Environmental Education* 2, no. 4 (1971): 38–42.

Tiffany, Paul. "Corporate Management of the 'External Environment': Bethlehem Steel, Ivy Lee, and the Origins of Public Relations in the American Steel Industry." *Essays in Economic & Business History* 5, no. 1–18 (1987).

Timberlake, Lloyd. "Catalyzing Change: A Short History of the WBCSD." Geneva: WBCSD, 2006.

Turl, Adam. "The Miners' Strike of 1977–78: Resisting the Employers' Offensive." *International Socialist Review*, November 2010.

Turner, George Kibbe. "Manufacturing Public Opinion: The New Art of Making Presidents by Press Bureau." *McClure's Magazine*, New York, July 1912. https://babel.hathitrust.org/cgi/pt?id=msu.31293006586253&view=1up&seq=7.

Turner, Frederick Jackson. "The Significance of the Frontier in American History." In *The Frontier in American History*. New York: Henry Holt, 1920.

"Two Views: National Environmental Development Association." *Environment: Science and Policy for Sustainable Development* 23, no. 6 (1981): 17–20.

Uldam, Julie. "Activism and the Online Mediation Opportunity Structure: Attempts to Impact Global Climate Change Policies?" *Policy and Internet* 5, no. 1 (2013): 56–75.

"Union Camp, Georgia Pacific, and Dravo Donate Key Natural Areas." Natural Assets: A Report of Corporate Achievements in Land Conservation. *The Nature Conservancy* 1, no. 1 (1976): 1–2.

Union of Concerned Scientists. "Smoke, Mirrors & Hot Air: How ExxonMobil Uses Big Tobacco's Tactics to Manufacture Uncertainty on Climate Science."

Cambridge, MA: Union of Concerned Scientists, 2007. papers3://publication/uuid/ E477490C-E46D-4E14-B6EA-9602C9B9E49B.

United Nations. *An Action Plan for the Human Environment.* Report by the Secretary-General. Conference on the Human Environment, Stockholm, 5-16 June 1972. Provisional Agenda Item 16. A/CONF.48/5. 9 February 1972.

United Nations Conference on Environment and Development. *Agenda 21: Programme of Action for Sustainable Development; Rio Declaration on Environment and Development; Statement of Forest Principles.* New York: United Nations, Department of Public Information, 1993.

United Nations Environment Programme. "World Industry Conference on Environmental Management (WICEM): Outcome and Reactions." Paris: UNEP, Industry and Environment Office, 1984.

United States Department of the Interior National Park Service. "National Historic Landmark Nomination: Ludlow Tent Colony Site." Washington, DC: United States Department of the Interior National Park Service, 2008.

US Congress. "A Resolution to Investigate Violations of the Right of Free Speech and Assembly and Interference with the Right of Labor to Organize and Bargain Collectively." 76th Congress, 1st Session (Part 37: Supplementary Exhibits), 16 January 1939.

US Congress. "To Amend and Extend Authorizations for the Federal Water Pollution Control Act." 95th Congress, 1st Session, 1 March 1977.

US Congress. "An Act Relating to Rights of Way through Certain Parks, Reservations, and Other Public Lands." 56th Congress, 2nd Session, 15 February 1901.

US Congress. "Congressional Record 51." 63rd Congress, 2nd Session, 5 May 1914.

US Congress. "Pinchot (Forest Service, U.S. Department of Agriculture) to Hon. Charles F. Scott (Chairman, Committee on Agriculture, House of Representatives)." 60th Congress, 1st Session, 30 March 1908.

US Congress. "Hearings before the Committee on Agriculture, Agricultural Appropriations Bill." 60th Congress, 2nd Session, 23 January 1908.

US Congress. "Maintenance of a Lobby to Influence Legislation." 63rd Congress, 1st Session, 25 June 1913.

US Congress. "National Coal Policy Project: Hearing before the Subcommittee on Energy and Power." 95th Congress, 2nd Session, 10 April 1978.

US Congress. "National Industrial Recovery Act." 73rd Congress, 1st Session, 16 June 1933.

US Congress. "Statement of Thomas A. Young—Clean Air Act Oversight." 93rd Congress, 1st Session, September 1973.

Useem, Bert, and Mayer N. Zald. "From Pressure Group to Social Movement: Organizational Dilemmas of the Effort to Promote Nuclear Power." *Social Problems* 30, no. 2 (1982): 144–156.

Useem, Michael. *The Inner Circle: Large Corporations and the Rise of Business Political Activity in the U.S. and U.K.* Oxford: Oxford University Press, 1984.

Van Heuverswyn, A., and J. Schuybroek. "Lobbying: An Old Profession Rediscovered." *Gestion 2000 (Louvain)* 5 (1990): 131–141.

Vance, Arthur T. "The Value of Publicity in Reform." *Annals of the American Academy of Political and Social Science* 29 (1907): 87–92.

Vesty, Gillian Maree, Abby Telgenkamp, & Philip J. Roscoe. "Creating Numbers: Carbon and Capital Investment." *Accounting, Auditing & Accountability Journal* 28, no.3 (2015): 302–324.

Vietor, Richard H. K. *Environmental Politics and the Coal Coalition.* College Station: Texas A&M University Press, 1980.

Vogel, David. *Fluctuating Fortunes: The Political Power of Business in America.* Washington, DC: Beard Books, 1989.

Vogel, David. "The Public-Interest Movement and the American Reform Tradition." *Political Science Quarterly* 95, no. 4 (Winter 1980–1981): 607–627.

Walker, Edward T. *Grassroots for Hire: Public Affairs Consultants in American Democracy.* New York: Cambridge University Press, 2014.

Walker, Edward T. "Legitimating the Corporation through Public Participation." In *Democratizing Inequalities: Dilemmas of the New Public Participation,* ed. Caroline W. Lee, Michael McQuarrie, and Edward T. Walker, 66–80. New York: New York University Press, 2015.

Warner, Michael. *Publics and Counterpublics.* New York: Zone Books, 2005.

Warren, Kenneth. *The American Steel Industry, 1850–1970: A Geographical Interpretation.* Pittsburgh: University of Pittsburgh Press, 1987.

Waterhouse, Benjamin. *Lobbying America: The Politics of Business from Nixon to NAFTA.* Princeton, NJ: Princeton University Press, 2014.

"We Can Work with You." Nature Conservancy Case Study, Silver Anvil Award, Public Relations Society of America, PRSA Box 126-13, 1976.

Wessel, Milton R. *Science and Conscience.* New York: Columbia University Press, 1980.

Wessel, Milton R. *The Rule of Reason: A New Approach to Corporate Litigation.* Reading, MA: Addison-Wesley, 1976.

Weston, William W. "Public Relations: Trustee of a Free Society." *Public Relations Review* 1, no. 2 (1975): 5–14.

"What Is the N.A.M.?" New York: National Association of Manufacturers, 1944.

White, Ahmed. *The Last Great Strike: Little Steel, the CIO, and the Struggle for Labor Rights in New Deal America.* Berkeley: University of California Press, 2016.

White, Jenny, and Lisa A. Bero. "Corporate Manipulation of Research: Strategies Are Similar Across Five Industries." *Stanford Law and Policy Review* 21, no. 1 (2010): 105–134.

Whiteside, Thomas. "Profiles: A Countervailing Force- I." *New Yorker,* October 1973.

Whyte, William G. "Remarks of William G. Whyte Before Public Relations Society of America." Honolulu, Hawaii. PRSA Box 171, 15 November 1973.

Williams, Amanda, Gail Whiteman, and John N. Parker. "Backstage Interorganizational Collaboration: Corporate Endorsement of the Sustainable Development Goals." *Academy of Management Discoveries* 5, no. 4 (2019): 367–395.

Williams, Juan. "Return from the Nadir." *Washington Post,* 23 May 1982.

Wittenberg, Ernest. "In Brussels, a 'Gucci Gulch.'" *New York Times,* 24 March 1989.

Wittenberg, Ernest. "How Lobbying Helps Make Democracy Work: A Speech Delivered to the Brazilian Public Relations Congress in Brasilia on 2 September 1982." *Vital Speeches of the Day* 49, no. 2 (1 November 1982).

Wittenberg, Ernest, and Elisabeth Wittenberg. *How to Win in Washington: Very Practical Advice about Lobbying, the Grassroots, and the Media.* Cambridge, MA: Basil Blackwell, 1989. https://www.industrydocumentslibrary.ucsf.edu/tobacco/docs/yqfl0051.

Wood, Peter. "Business-suited Saviors of Nation's Vanishing Wilds." *Smithsonian Magazine* (December 1978): 77–85.

Wood, Tim. "Corporate Front Groups and the Making of a Petro-Public." PhD dissertation, New York University, 2018.

Wood, Tim, and Melissa Aronczyk. "Publicity and Transparency." *American Behavioral Scientist* 64, no. 11 (2020): 1531–1544.

Wylie, Sara, Nicholas Shapiro, and Max Liboiron. "Making and Doing Politics Through Grassroots Scientific Research on the Energy and Petrochemical Industries." *Engaging Science, Technology, and Society* 3 (2017): 393–425.

Yakowitz, Jane. "Tragedy of the Data Commons." *Harvard Journal of Law & Technology* 25, no. 1 (2011): 1–67.

Yates, JoAnne. "Creating Organizational Memory: Systematic Management and Internal Communication in Manufacturing Firms, 1880-1920." Working Paper #2006-88, Massachusetts Institute of Technology, Cambridge, MA, April 1988.

Yosie, Terry F. "Emerging Strategies to Manage System-Level Risks: An Examination of Private Sector, Government and Non-Governmental Organization Initiatives." In *Improving Risk Regulation*, ed. International Risk Governance Council. Lausanne: IRGC, 2015: 27–41. https://irgc.org/risk-governance/risk-regulation/improving-risk-regulation/.

Yosie, Terry F., and Timothy D. Herbst. "Using Stakeholder Processes in Environmental Decisionmaking: An Evaluation of Lessons Learned, Key Issues, and Future Challenges." 1998. https://www.gdrc.org/decision/nr98ab01.pdf.

Zagorin, Adam. "The Euro Peddler (De Euroleurder)." *Elsevier Weekly Magazine* (in Dutch), June 1989.

Zuboff, Shoshana. "Big Other: Surveillance Capitalism and the Prospects of an Information Civilization." *Journal of Information Technology* 30, no. 1 (2015): 75–89.

Zuboff, Shoshana. "Surveillance Capitalism and the Challenge of Collective Action." *New Labor Forum* 28, no. 1 (2019): 10–29.

Index

Printed in the USA/Agawam, MA
May 2, 2022

792471.021